SOTHIS PRESS

Sothis Press aims to completely reimagine the public domain book with an emphasis on universal access to captivatingly designed print editions. We believe that technology will enable every single book ever published to be perpetually accessible in beautiful print form. Sothis is beauty in ubiquity.

www.sothis.press
1 Waterhouse Square, London, ECIN 2ST

Types of Aeroplanes

Instruction Paper

Charles B. Hayward

SOTHIS PRESS

ISBN 978-1-0365-2620-7

Types of Aeroplanes is
in the public domain.

TYPES OF AEROPLANES

PART I

INSTRUCTION PAPER

PREPARED BY

CHARLES B. HAYWARD

MEMBER, SOCIETY OF AUTOMOBILE ENGINEERS
MEMBER, THE AERONAUTICAL SOCIETY
FORMERLY SECRETARY, SOCIETY OF AUTOMOBILE ENGINEERS
FORMERLY ENGINEERING EDITOR, *The Automobile*

AMERICAN SCHOOL OF CORRESPONDENCE

CHICAGO ILLINOIS

U.S.A.

TYPES OF AEROPLANES

PART I

STANDARD TYPES

General Survey. In view of the fact that aeroplane design can hardly be said to have progressed beyond its inception, it may appear to be somewhat of a misnomer to refer to standard types. There are, however, a certain number of designs in biplanes and monoplanes, constructed according to well-defined models, and after which the majority of others are patterned. While these are more or less similar in their fundamental characteristics, they vary from one another in important details of size, arrangement, and efficiency of their parts. For the purpose of comparison, a discussion of their distinguishing features, as well as their merits and demerits, is appended. A study of this will be found of the greatest value as a means of obtaining a knowledge of the chief characteristics of the best-known aeroplanes.

The fifteen most prominent and distinctive types are described in detail, the order in which they are taken up not being based on any quality of the machines themselves. The biplanes are eight in number, as follows: Wright, Wright Racer (Baby), Curtiss, Voisin, New Model Voisin, Farman, Sommer, and Cody.

The monoplanes are seven in number, as follows: Antoinette, Santos-Dumont, Bleriot XI, Bleriot XII, Grade, Pelterie, and Pfitzner.

With few exceptions the machines in question as described in the following paragraphs have been flown thousands of miles and used over extended periods by a great number of aviators and amateurs, and they have likewise been copied in hundreds of other machines, but as the result of the experience thus gained, their builders have inaugurated various changes, not merely of dimensions, but of construction in some cases and of principle in others. Most of these changes have been brought about during 1911. In not a few instances, the changes have been of sufficient importance to warrant giving the new machine a new title. Wherever the changes have altered the machine materially, the details are given just after the description of the standard type.

In addition to the foregoing, there are some special types the distinguishing features of which merit reference. Many other types of successful biplanes and monoplanes are in use, but they differ so slightly from one or another of those described here that any detailed mention of them would only lead to confusion. The great number of machines now being built in this country by individual experimenters or by manufacturers are either replicas of those detailed or are modifications of them.

Nomenclature. Despite the phenomenally rapid development of aviation, its terminology has kept pace so that there are a number of expressions the meaning of which must be explained before attempting a description of the machines themselves.

SUPPORTING PLANE. By supporting plane is meant the main lifting surface as distinguished from all auxiliary or stabilizing surfaces.

DIRECTION AND ELEVATION RUDDERS. Direction rudder refers to the movable, vertical surface used for steering to the right or left, while the elevation rudder is a horizontal surface the function of which will be obvious.

TRANSVERSE CONTROL. Transverse control is the device employed for the preservation of lateral balance when flying straightaway and for maintaining an artificial inclination of the machine when rounding turns.

KEELS. Keels are fixed surfaces intended to aid in the preservation of stability; they exert neither lifting effect nor rudder action.

SPREAD. Spread is the maximum horizontal dimension perpendicular to the line of flight.

DEPTH. Depth is the dimension of the plane parallel to the line of flight.

ASPECT RATIO. By "aspect ratio" is meant the proportion of spread to depth and it constitutes a factor for defining the shape of the supporting plane.

For the purpose of more clearly showing the variation in size of the different types, detailed and dimensioned plans and elevations of each machine are given. Most of these are drawn to the same scale, thus enabling a direct, graphic comparison of the types. But it must be borne in mind, inasmuch as aviators are constantly changing and rechanging the dimensions of their machines, without recording such alterations, many of the dimensions given here are necessarily approximate. In all cases, however, the most recent and accurate data, as furnished by the large number of references consulted as well as by close personal inspection, has been employed.

BIPLANES

Wright. This Wright machine, Fig. 1, is the original Wright type of which many are made and used in England, France, and Germany, there being Wright companies in those countries devoted to their manufacture and exploitation. The more recent Wright machines do not require a rail or weight for starting and the front elevation rudder has been discarded. Among the biplanes the Wright is almost twice as efficient as any other type, this being ascribed by French writers, particularly Berget, to the fact that a great deal of weight is saved by the starting device. This is what the latter was originally adopted for, but as no increase in power was found necessary when it was discarded for the four-wheeled chassis now employed, the contention does not hold good. In view of its much rougher construction as compared with the finely finished French machines, its efficiency is extremely high, owing in large measure, doubtless, to the employment of two propellers revolving at a comparatively slow speed.

Frame. Clear spruce and ash are used throughout in the construction of the frame, which is very simply but solidly built. The bracing wires are steel and are made to fit exactly, while the struts or separators are of elliptical form with the small edge facing the direction of motion. These struts are equipped with hooks at each end fitting in rings in the frames of the two planes. All exposed parts of the machine are painted with an aluminum mixture.

Supporting Planes. Two identical and superposed surfaces of canvas (fine, closely woven duck) stretched over and under wood ribs of light but strong built-up construction support the machine in the air. These surfaces, or planes, are 3 inches thick near the center and have a somewhat flatter and more regular curve than that commonly employed. The planes, which are spaced 6 feet apart, have a spread of 41 feet, a depth of 6.56 feet, and a total area of 538 square feet.

Elevation Rudder. In the Wright biplane the rudder is so constructed that when elevated it is automatically warped concavely on the under side, and when depressed it is curved in the opposite way. This materially adds to the force exerted. It is double surfaced, constituting a small biplane itself and has 70 square

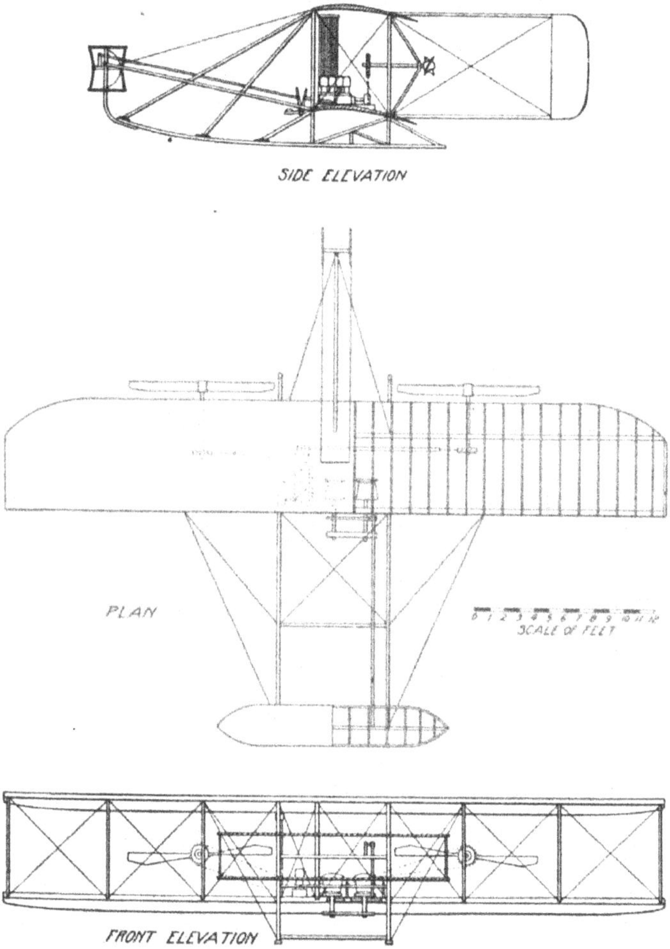

Fig. 1. Original Type of Wright Biplane

feet of area; it is placed well forward of the main planes, being supported on an extension of the landing skids. This rudder is controlled by a lever worked by the operator's left hand. To rise, the aviator pulls the lever toward him. This motion, transmitted to the rudder mechanism by a long, wood connecting rod, causes the rudder to turn upward relative to the line of flight and consequently the machine rises. Reversing the movement causes it to descend.

Direction Rudder. The direction rudder is placed in the rear on the center line, and consists of two identical and parallel vertical surfaces with a total area of 23 square feet. It is governed by the right-hand lever, turning to the left being accomplished by pushing out and to the right by pulling in on it. The control is not employed exactly in this manner, however, as a sidewise movement of the same lever also serves to warp the planes—a feature indispensable to lateral equilibrium in rounding turns. The two motions of the lever are very intimately connected in their effect upon the control.

Transverse Control. Transverse control is the famous warping device invented by the Wrights for the preservation of lateral balance and for artificial inclination in making turns, and is employed in a similar or modified form in almost every aeroplane thus far constructed, the Pfitzner monoplane constituting the most radical departure from it. To permit of this warping, the rear vertical panel of the main cell, or double plane, is divided into three sections. The central panel is solidly braced and extends on either side of the center to the second strut from each end. From these struts, the rear horizontal crosspieces are merely hinged instead of being continued portions of the crosspiece at the center, and the two vertical panels on either end are not cross braced. These two rear end sections of the cell are, therefore, movable vertically. The entire front of the machine, as well as the ribs inside the supporting planes, however, are perfectly rigid, there being no helical torsion of the ribs themselves, as commonly supposed. Cables connect these two sections of the planes together and lead to the right-hand lever. The operation is as follows: If the machine suddenly tilts or dips down, at the right end, for example, the lever is moved to the left. This action pulls down the rear right ends of the surfaces and at the same time pulls the left ends upward. An increase in the incident angle of the outer end of the plane on the depressed side and a decrease of the incident angle on the opposite side, are thus brought about, righting the machine at once. During this operation, the entire front face of the cell as well as the rear central section remain perfectly rigid in every sense.

The warping apparatus is also interconnected with the direction rudder and the simultaneous action of both is depended upon. This is one of the chief claims of the original Wright patent, and in actual practice the direction rudder and transverse control of the machine are rarely, if ever, worked separately. To make a turn to the left, for example, it is evident that if this same lever is moved in an arc, outward and to the left, somewhat similar to

Fig. 2. Tail of Short Wright Biplane Showing Addition of Horizontal Keel at Rear

the contour of the desired turn, not only will the surfaces be warped so as to raise the right end, but the direction rudder is also set to give the desired change of travel, and the combined action of the two is prompt and very effective.

There are no keels on the original Wright biplane, but since the elimination of the forward elevating rudder, these have been introduced in the later type, Fig. 2. In the older machine, a small, pivoted, vertical surface is placed in front to indicate any change in direction of the relative air current.

Power Plant. The power plant consists of a four-cylinder, vertical, four-cycle, water-cooled motor built by the Wrights themselves and rated at 25 to 28 horse-power, which drives two double-bladed propellers in opposite directions by chains and sprockets. The propellers are of laminated wood construction, made of clear spruce, measuring 8.5 feet in diameter and having a 9-foot pitch. They rotate at 400 r.p.m., or only about one third the speed at which the usual single propeller is ordinarily driven, and are placed at the rear of the main cell, at equal distances on either side of the center.

Running Gear. As already mentioned, the original mounting was on skids only, but since about July, 1910, all of the Wright

Fig. 3. Brookins in Headless Wright Just About to Leave the Ground

machines have been fitted with four pneumatic-tired wheels attached to a rectangular frame. The total weight of the machine described above is 1,050 to 1,150 pounds and the speed 40 miles per hour; 41 pounds are lifted per horse-power of the motor and 2.05 pounds per square foot of supporting surface. The aspect ratio is 6.25 to 1. These figures, however, apply only to this particular machine, as the Wright biplane built for the United States Signal Corps, as well as those constructed by the Aerial Company of France, have a spread of only 36 feet with a total supporting surface of 490 square feet.

French Wright. In the French Wright machines, the aviator sits next to the motor, and when instructing Count Lambert and

M. Tissandier in the winter of 1909 at Pau, Wilbur Wright had fitted to the machine an extra lever to control the elevation rudder on the right side of the passenger who sat next to the motor. The position of the levers for the passenger was, therefore, the reverse of the usual one. Messrs. Tissandier and Lambert, having learned to operate in this manner, have never changed, but as they in turn have become the instructors of many purchasers of Wright machines, their pupils are taught to control in the normal manner.

New Model Wright. The new Wright machine, introduced in the summer of 1910 and first seen in public at the Asbury Park Meet, has no front elevation rudder and was, therefore, popularly dubbed the "headless" Wright, shown in Figs. 3 and 4. The elevation of the machine is controlled by the rear horizontal surface alone. This machine is also smaller and faster, its spread being 39 feet, depth 5.5 feet, and supporting surface 410 square feet. With the 30-horse-power motor employed, the lift is 37 pounds per horse-power, or 2.5 pounds per square foot of surface. The aspect ratio is 7.1 to 1.

Fig. 4. Headless Wright in Flight

Wright Racer. The Wright Racer is officially known as "Model R" by the manufacturers, but owing to its diminutive size was immediately christened the "Baby Wright" on its first appearance at the International Meet at Belmont Park in 1910. It was especially designed for high speed and one of this model with an eight-cylinder motor was entered in the Gordon-Bennett cup race, but owing to an accident it did not take part. This machine is shown in Fig. 5 with Orville Wright driving, Hoxsey holding the machine on the right, and Brookins on the left. As the engine is running, the propellers do not show in the illustration. Sufficient accommodation only for the aviator is provided so that it is a one-man machine. It is said to be the fastest climbing aeroplane ever built,

Johnstone's record of 9,714 feet made at Belmont Park having been accomplished on this model.

Frame. This machine is of the same headless type as that brought out in the larger size during the early part of 1910. The construction of the frame throughout is the same as in the latter.

Supporting Planes. The supporting planes are of the same design and construction as in the larger machine, but they have a spread of only 26½ feet by a depth of 3 feet 7 inches, giving a total area of but slightly over 185 square feet. The length fore and aft is 24 feet, while the height from the ground to the top of the upper plane is but 6 feet 10 inches.

Elevation Rudder. The elevation rudder, as well as the direction rudder, is of the same design, construction, and operation as the

Fig. 5. Wright Baby Racer, with Orville Wright at the Wheel

standard Wright flyer, the dimensions merely being made to correspond to its smaller size.

Transverse Control. The regular Wright warping device in connection with the control of the direction rudder is employed, as in the larger machines.

Power Plant. The power plant is an eight-cylinder, V-type, 50- to 60-horse-power motor which is characterized by the same features of design as the standard Wright four-cylinder motor used on the larger machines. It drives two two-bladed wood propellers in opposite directions through the medium of chains and sprockets, and, so far as may be noted by a casual examination, they are identically the same as those employed on the regular Wright machines and are designed to run at the same speed, *i.e.*, about 400 r.p.m., the speed of the motor being 1,300 r.p.m.

General. The seat for the aviator is directly in front of and in line with the motor and there is no provision for carrying a passenger, owing to the extremely small size of the machine. It is, in fact, a "fly-about," to coin a term analogous to that prevalent in the automobile field. The machine is mounted on two pairs of pneumatic-tired wheels straddling each of the skids and placed directly under the center of the machine.

The weight of the machine alone is only 585 pounds, its total weight in flight ranging from 735 to 800 pounds, thus lifting 13.3 pounds per horse-power, taking as a basis the maximum weight of 800 pounds and putting the horse-power of the motor down as 60. On the same basis of total weight, the loading is 4.27 pounds per square foot of surface. The aspect ratio is 7.4 to 1.

Wright Model B. In automobile parlance, this is the standard 1912 Wright Model, and while it shows few or no departures from the principles already established in its predecessors, it is distinguished by a number of refinements. The spread is 39 feet and the chord 6 feet 2 inches, the main planes being built in three sections and covered with Goodyear rubberized fabric in place of the canvas formerly employed. The fabric is laid diagonally and is attached to each section independently, the sections being laced together when the machine is assembled. The main spars are of spruce, as is most of the rest of the woodwork, $1\frac{3}{4} \times 1\frac{1}{4}$ inches, the greatest dimension being vertical in the front spar and horizontal in the rear spar. They are larger in the middle section of the lower plane, ash being used in the rear of the latter. There are 34 ribs to each plane, spaced a foot apart in the center and wider toward the lateral extremities of the planes. The ribs which come near struts are solid between the main spars, the others being built up of an upper and lower strip with blocks spaced about six inches as distance pieces. The two ribs that support the engine and the two seat ribs are the only ones between the spars of the lower main plane in its center section. There are nine pairs of uprights of various sizes, the outer two sets on each end being secured to the planes by the familiar flexible joint, the remainder having a form of socket joint. A few turnbuckles have made their appearance in the center section, doubtless to facilitate replacement of the engine or other parts. All the steel piano wires not fitted with turnbuckles are cut to length and are interchangeable. When setting up the planes, the wires are attached and the struts then sprung into place. These guy wires are cut and the loop bent by a special machine at the factory. As the wire employed has a breaking strength of 800

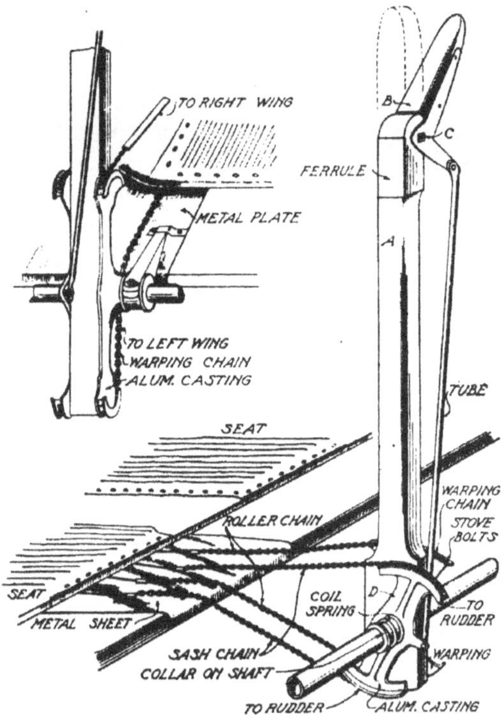

Fig. 6. Details of Wright Model B Combination Warping and Direction Lever

to 2,400 pounds, according to size, there should be no occasion for adjustment on account of stretch. The curve of the planes is 1 in 20, the greatest depth being two fifths of the chord back from the front edge. The aspect ratio is 6.25 to 1.

The small semicircular fins or "blinkers" familiar on the 1910 machine have given place to two sets on the latest machines, due to the fact that greater area is required as the skids have been shortened, thus bringing these surfaces closer to the main planes. Their shape is that of small jibs.

The vertical rudder is, in general, of the same construction as in the earlier models, though somewhat smaller. The rudder is operated by the combination warping and direction lever, Fig. 6. As shown, this lever also warps the wings. By "breaking" the top section *B*, either to the left or to the right (without moving the rest of the lever from its position), the rudder is moved only to steer left or right, respectively. In making flat turns, without banking, the top section only of the lever is used. The movement is entirely a natural or instinctive one. This separate movement of the rudder is obtained by having the sector *D*, movably mounted, capable of individual action with respect to lever section *A*, through the steel tube actuated by the section *B* of the lever. The wire which goes over the top of sector *D* must go to the left side of the rudder cross bar.

The front third of the elevator surface is held rigid while the remainder is flexible. This is operated by a forward and backward movement of the elevator lever, the wires being crossed so that pushing out on the lever steers down and pulling toward the operator causes the machine to ascend. The cloth is laid on diagonally and only one surface is used, the ribs and spars running through pockets in the cloth. There is a second elevator lever which can be used by a student passenger, who would then work the warping lever (and rudder) with his right hand. Some of the Wright aviators use the seat next to the engine with the warping lever at the left, while others sit on the outside seat. This second elevator lever has a disk attached, encompassed on its periphery by a flat steel friction band to hold the lever in any set position.

While the control of the machine does not appear to be instinctive, it certainly is very easy to learn and, after having it once impressed upon the mind, is very satisfactory. It would seem that the exertion of moving the warping lever fore and aft is a great deal less than if it were arranged to move sideways as in some other machines. The warping is effected by the lever *A*, Fig. 6. Pushing forward raises the left wing and depresses the right; the same movement turns the rudder to the left—besides having a lesser angle of incidence, when the lever as a whole is used. The wiring for the warping is shown in the diagrammatic sketch, Fig. 7. The rear spars of the two end sections of the planes are hinged to those of the center section, so that warping may be accomplished without flexing the spar. The lever arrangements have varied on many of the machines. Some are flown with the aviator using the left hand for warping. Students taught by these use the right hand for warping, as a rule, and this is now the practice in "breaking in" flyers in order that any passenger or other weight they may carry will occupy a central position on the machine and retain the balance. However, one or two machines have been put out with two warping and two elevating levers, for those who desire to fly together, both having learned the use of the same hand for warping.

Referring to the combination warping and rudder lever, Fig. 6, the lever *A* is jointed or hinged at the top. The short section *B*

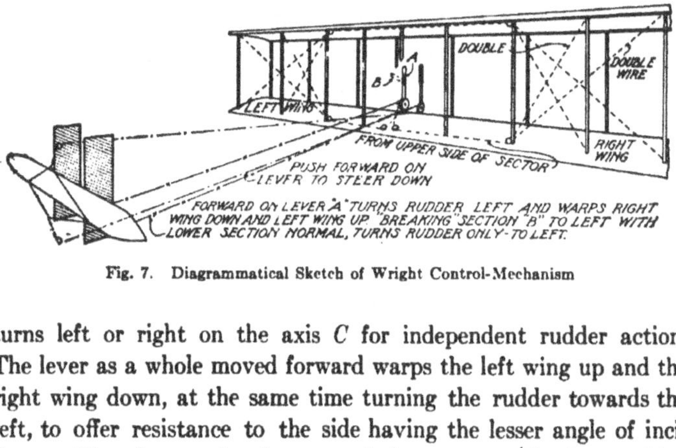

Fig. 7. Diagrammatical Sketch of Wright Control-Mechanism

turns left or right on the axis *C* for independent rudder action. The lever as a whole moved forward warps the left wing up and the right wing down, at the same time turning the rudder towards the left, to offer resistance to the side having the lesser angle of incidence. The elevator is also warped down to enable the machine to gain speed, and the aeroplane has begun to bank, the right side being the higher. Next, this combination lever as a whole is gradually brought back to normal position, as the aeroplane is now at almost a forty-five degree angle. At this stage with this lever (as one) normal, and the wings straightened out, the top section of the lever is "broken" over to the left, which turns the rudder only to this side. This operation is gone through in making short circles, or spirals, for which the Wright machine is famous. For right spirals, the reverse of the operation just described must be carried out, care being taken to straighten out before the machine has banked at so steep an angle as to make recovery impossible. In Fig. 6 the section *B* is broken to the left, turning the rudder only in that direction.

The motor on this machine does not differ except in a few details from that which the Wright Brothers have been building for their own machines ever since they began flying. One of the innovations consists of an emergency shut-off of the power, consisting of a wire conveniently placed over the aviator's head. Pulling this raises the exhaust valves and thus cuts off the power of the motor, without bringing it to a sudden and dead stop, as in the case where the switch for short-circuiting the Mea magneto is closed. The power can thus be cut down considerably without bringing the motor to a stop. The same method of feeding the gasoline directly to the inlet manifold by means of a gear pump, and without a carbureter, is still retained. As its speed increases with that of the engine, the amount of fuel fed is always in proportion to the latter's speed. Retarding or advancing the spark is accordingly the only method of controlling the speed of the motor, apart from the exhaust valve control previously mentioned. A pedal in front of the aviator sets the spark back to facilitate safe starting of the motor, and the magneto is provided with a catch to hold it in the retarded position, so that an aviator may start his own machine without danger of having it run away from him before he can get into the seat. The weight of the bare engine is 180 pounds and it consumes about 4 gallons of gasoline per hour, the 12-gallon tank accordingly providing sufficient for a three-hour flight.

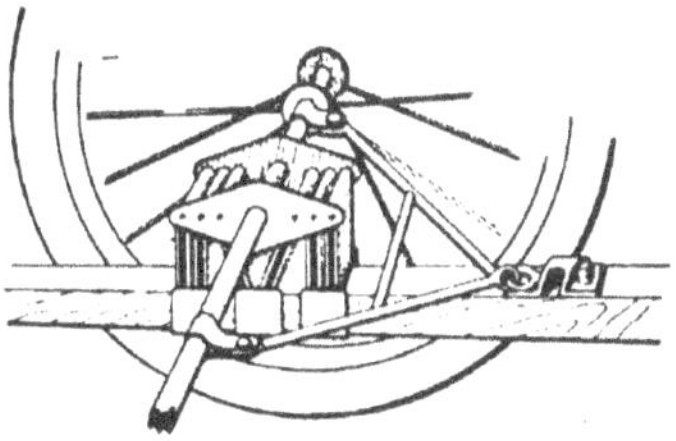

Fig. 8. Wheel Mounting Details, Wright Model B

The engine is mounted at either end of the base on cross members, which in turn rest on the solid engine foundation ribs. Duplicate sprockets, which are screwed and locked to the crank shaft back of the flywheel, drive by means of special roller chains the

nust be carried
e machine has
impossible. In
he rudder only

cept in a few
n building for
e of the inner
er, consisting
Pulling this
of the motor,
e case where
losed. The
ringing the
ine directly
by means
without a
ained. As
h that of
it of fuel
portion to
arding or
s accord-
of con-
e motor,
ned. A
ate safe
a catch
y start
m him
gine is
hour,
e-hour

mem-
Dupli-
shaft
s the

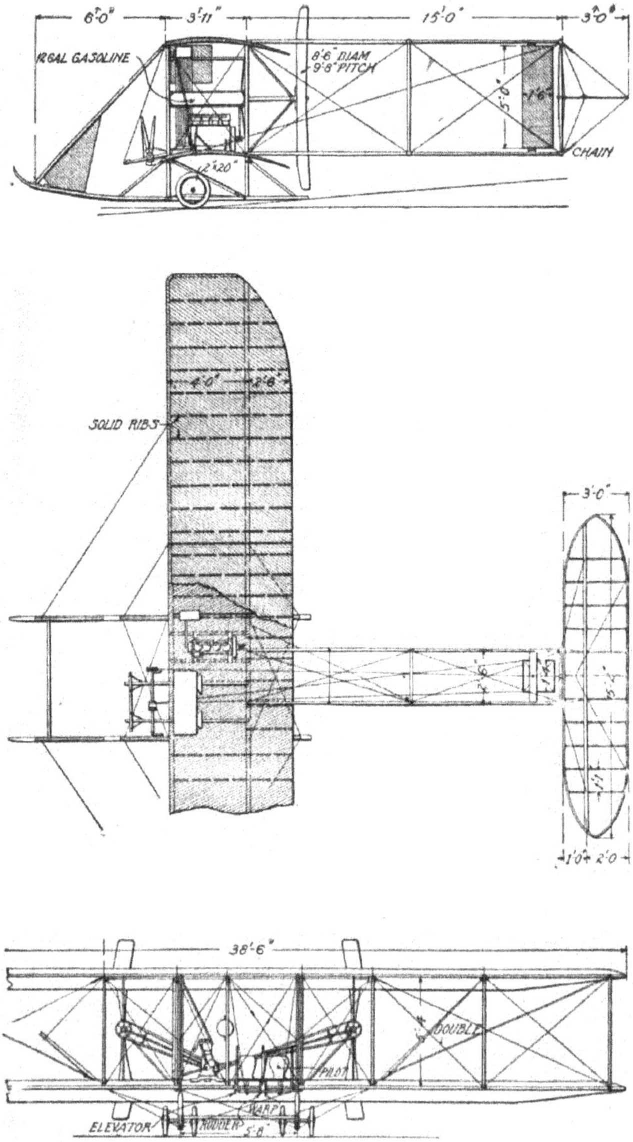

Fig. 9. Detailed Diagrams of Wright Model B

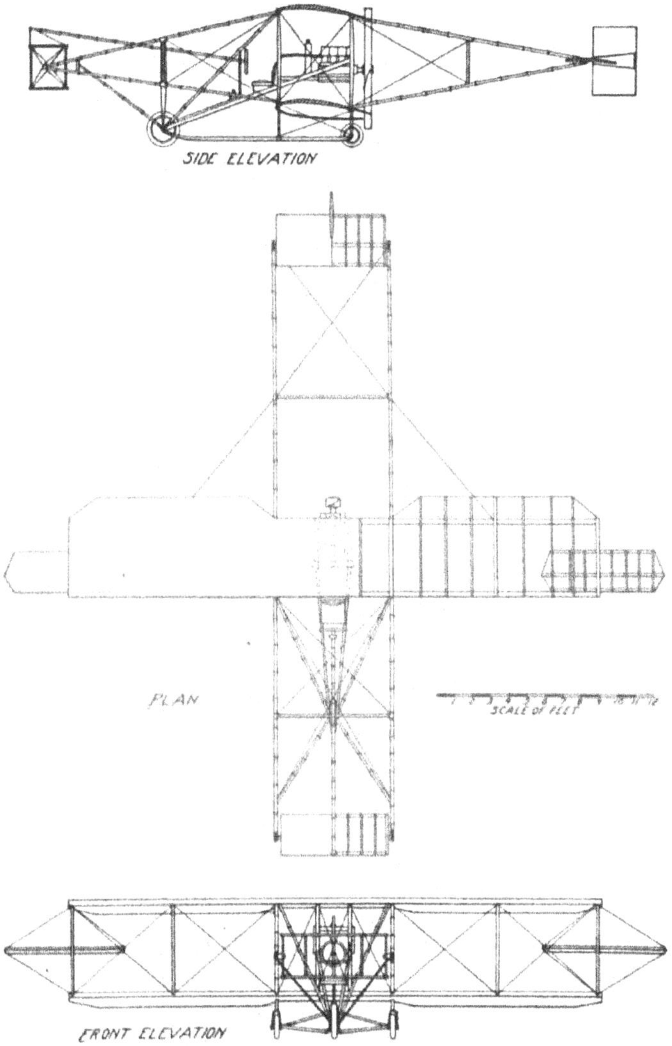

Fig. 10. Detailed Diagram of Curtiss Biplane

two propellers, their speed being geared down in the ratio of 11 to 34. At an engine speed of 1,325 r.p.m., the propellers turn at

Fig. 11. Curtiss on His Trip from Albany to New York City. Leaving Poughkeepsie

428 r.p.m., giving a flying thrust of about 250 pounds. Adjustable stays are provided for tightening the chains.

For the landing gear, wheels are used in combination with the usual skid arrangement, the skids themselves having been very much shortened. The method of mounting the wheels is illustrated in Fig. 8. The complete machine is illustrated in Fig. 9.

With operator and passenger, ready to fly, the machine weighs about 1,250 pounds. The weight thus carried per horse-power is about 40 pounds, while the loading on the above basis figures out at but $2\frac{1}{2}$ pounds per square foot. Lancaster gives the Wright machine an efficiency of 63 per cent, after deducting 5 per cent for loss in the chains. In a book by Eiffel (1911), it is stated that 30

Fig. 12. Curtiss on His Albany to New York Trip. Flying Down the Hudson

horse-power is required to fly the Wright machine, which, in view of the facts, is obviously an erroneous conclusion.

Curtiss. The Curtiss biplane, Fig. 10, embodies in its construction several features that distinguished the aeroplanes built by the Aerial Experiment Association, of which Glenn H. Curtiss was a member. The first flight of this type was made in June, 1909. At Rheims, France, in August of the same year, this miniature biplane captured the Gordon-Bennett trophy as well as several other prizes, under the able guidance of Curtiss. A number of these machines are being flown and have a great many estimable performances to their credit, such as the flight of Curtiss from Albany to New York, illustrated in Figs. 11, 12, and 13, and Ely's flight from the deck of a man-of-war to the shore and back. The Curtiss is one of the fastest biplanes in use.

Fig. 13. Curtiss Rounding the Statue of Liberty

Frame. The main cell and smaller parts are made of ash and spruce, while the long outriggers are of bamboo, several of the members of the frame meeting at the front wheel of the landing chassis. Small steel cables and wires are employed for bracing.

Supporting Planes. The supporting planes consist of two identical directly-superposed surfaces made of one layer each of Baldwin rubber silk tacked to spruce ribs and laced to the frame, and are of highly-finished construction. A distance of 5 feet separates the two surfaces. Their spread is 26.42 feet, depth 4.5 feet, and total area 220 square feet.

Elevation Rudder. The elevation rudder is a small biplane cell having two similar surfaces of a total area of 24 square feet and mounted on bamboo outriggers on the meeting point of which it is pivoted, Fig. 14. It is controlled by a long, bamboo pole attached to the stanchion on which the steering wheel is mounted. To descend, the operator pushes out on the wheel, and to ascend draws it toward him. In Fig. 14, Curtiss is shown at the wheel.

Direction Rudder. For steering to right or left, a single, vertical surface of 6.6 square feet of area is pivoted at the meeting point of a similar pair of bamboo outriggers extending to the rear. It is operated from the steering wheel by cables running through the hollow outriggers.

Transverse Control. Transverse control consists of two independent balancing planes, or ailerons, of 12 square feet area each, which are shown very clearly in Fig. 15. They are placed at each end of the main cell and are pivoted midway between the upper and lower main planes. They are designed to preserve the lateral balance and are tipped inversely by means of a brace fitted to the

Fig. 14. Elevation Rudder and Steering Gear on Curtiss Machine

aviator's shoulders and controlled by the movement of his body. If the machine is depressed on the left side, the aviator leans to the right and in doing so shifts the brace, causing the aileron on the left side to turn down and the one on the right to turn up, the two being interconnected by cables, thus righting the machine. By "turning down" in this connection is meant a motion relative to the axis of the aileron itself and not to the line of flight. In other words, it swings on its supporting shaft. When turned down, its incidence, *i. e.*, the angle it makes with the line of flight, is positive and it therefore exerts a greater lifting force. When making a turn to the right, for instance, the aviator by leaning to the right, thus causing the left end to lift, can make a much sharper turn than with the use of the direction rudder alone, while the lateral balance is also preserved.

Keel. A horizontal fixed surface, or keel, is placed in the rear and has an important steadying effect. Its area is 15 square feet. A small, triangular vertical plane is sometimes placed in front and aids in turning.

Power Plant. In the original machine the power plant consisted of a four-cylinder, vertical, four-cycle, air-cooled motor of 25 horse-power, placed well up between the two main planes at the rear. It drives a two-bladed wood propeller direct at 1,200 r. p. m. The

Fig. 15. Curtiss Machine in Flight, Showing Ailerons and Position of Operator

propeller has a diameter of 6 feet and a pitch of 5 feet. In more recent Curtiss machines, an eight-cylinder, V-type, four-cycle, air-cooled motor of 50 horse-power is employed.

General. The seat for the aviator is on the framing in front of the main cell and in line with the motor, Fig. 15. When a passenger is carried, a seat is provided at one side and somewhat below the aviator. The machine runs on three pneumatic-tired wheels, rigidly fixed to the frame, no springs being provided. The total weight is from 530 to 570 pounds, and the speed is 47 m. p. h.; 22 pounds are lifted per horse-power, and 2.5 pounds per square foot of surface. The aspect ratio is 5.65 to 1.

During 1910, Willard, one of the Curtiss aviators, employed a much larger machine of exactly the same type, in which he succeeded in carrying three passengers besides himself. The supporting planes of this machine have a spread of 32 feet, a depth of 5 feet, and a total area of 316 square feet. The elevation rudder is 31 square feet in area, and the direction rudder 7.5 square feet, while the rear horizontal keel has an area of 17.5 square feet and the ailerons are each of 27 square feet area. A Curtiss eight-cylinder, 50-horse-power motor is employed to directly drive a 7-foot propeller at 1,100 r. p. m. The maximum total weight in flight is 1,150 pounds, thus lifting 22.6 pounds per horse-power, and 3.64 pounds per square foot of surface. The aspect ratio is 6.4 to 1. It was in a machine of this type that Curtiss made his flight from Albany to New York.

At the International Meet at Belmont Park, New York, in October, 1910, Curtiss exhibited a radically different type of machine. (See Fig. 49.) This embodied most of the constructional features already described, but instead of two similar planes there was one very large main surface with an extremely small superposed plane directly above the center of the latter. Though termed a biplane, it was practically a monoplane in everything but name. No opportunity was afforded of seeing what it could do in flight.

In a later type of the Curtiss, the ailerons are pivoted from the rear struts instead of the front ones, this doing away with their interference of the lifting power of the upper main plane. Head resistance has been cut down by double surfacing the planes, thus enclosing the ribs and beams, and also by adopting a single surface in place of the former biplane elevator. The axis of the new elevator is placed only 6 feet 9 inches in front of the main planes and has two short stays of bamboo between the wheel and the elevator instead of the elaborate and complicated structure formerly employed for staying. The rear tail flaps work in conjunction with the elevator, being pivoted at a point about 13 feet to the rear of the main planes. Two triangular stabilizing fins are used instead of the usual plane, their angle of incidence being about 2 degrees, which can readily be changed. The vertical rudder is placed between these two flaps and is pivoted back of its front edge, and it is operated by a tiller post or forward extension, instead of attaching the cables directly to the rudder itself. The span is 30 feet, chord 4 feet 2 inches, and the planes are 1 feet 5 inches apart vertically, the camber apparently not having been changed. The dimensions of the front elevator are 2 feet × 6 feet $3\frac{1}{2}$ inches, with a triangular vertical fin attached to it above. A standard eight-cylinder, V-type, 50-horse-power motor of Curtiss make drives a Curtiss two-bladed propeller, the laminated engine base being supported at the rear by steel tubing, which is also used to brace the entire rear section. In front, the base is bolted to two short laminated struts. The height of the base is 14 inches, and above this a triangle of 1-inch oval steel tubing extends to the top plane, where it is secured by a bolt. The

Fig. 16. Delagrange Model of Voisin Biplane

engine is placed 9 inches rearward from the rear beam, and the canvas seat is 8 inches forward of the front beam.

Voisin. The Voisin Brothers began their activity as constructors of aeroplanes as early as 1905, when they built gliders for both M. Archdeacon and M. Bleriot. These gliders were successfully operated over the Seine, being lifted from the surface of the river and towed at high speed by motorboats. In 1906, they built a power-driven machine after the designs of the late M. Delagrange, Fig. 16, and subsequently, after making a few changes in the design, built a machine for Henri Farman, Fig. 17, which was the first successful aeroplane of European manufacture. Since that time, the design of this type has remained substantially the same, except for the addition of several keels, Fig. 18. The Voisin biplane has been largely used abroad, over one hundred machines of this type having been built.

Frame. The frame is made of ash with steel joints and several parts of steel tubing. It consists essentially of a large box-cell

Fig. 17. Henri Farman in an Early Type of Voisin Biplane

mounted on a central chassis, while attached to it some distance in the rear is a smaller box-cell of the same form. This central chassis is really a unit in itself, carrying the wheels, the motor, the aviator's seat, and at the front the elevation rudder.

Supporting Planes. Two main supporting planes of similar dimensions and directly superposed are employed, the surfaces consisting of continental cloth (a cotton and rubber fabric) stretched

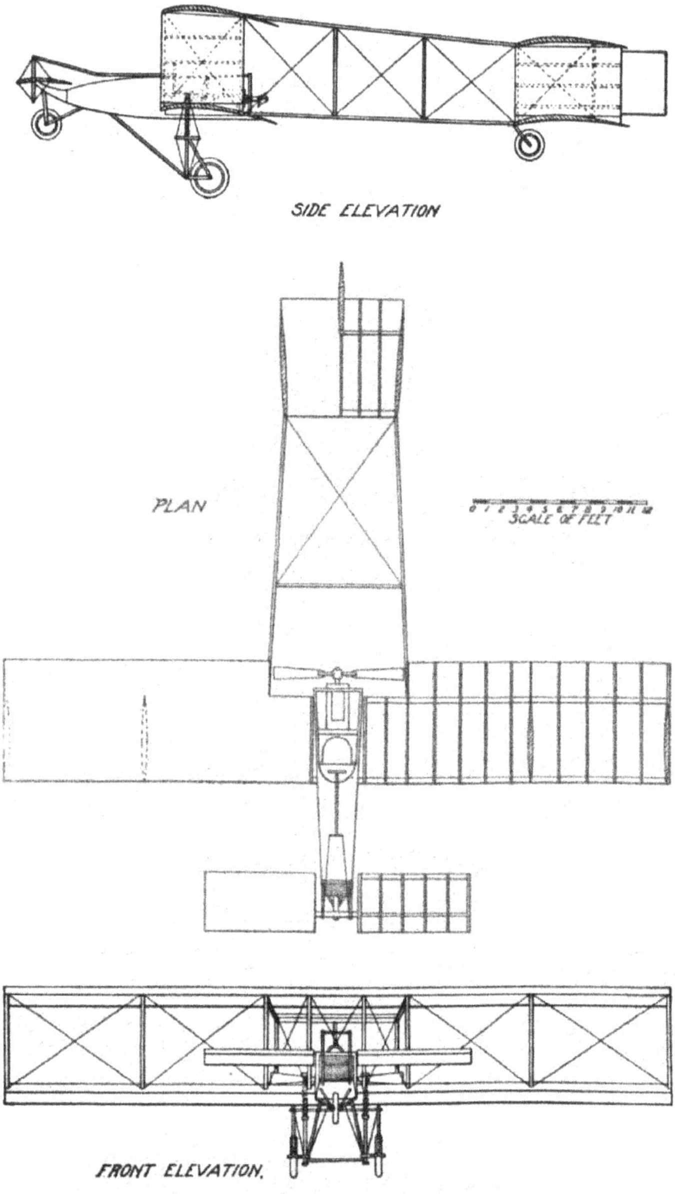

Fig. 18. Voisin Biplane with Vertical Keels

over ash ribs. Their form is rectangular. The spread is 37.8 feet, the depth 6.56 feet, and the total area 496 feet.

Direction Rudder. A single surface of 25 square feet area, placed in the center of the rear cells, is used for directing the machine. It is operated by means of a steering wheel and cables similar to those on a boat.

Elevation Rudder. The elevation rudder consists of a single surface of 41 square feet area situated at the forward end of the central chassis, and is controlled by a lever system attached to the axis of the steering wheel. By pushing out on the steering wheel, the rudder's inclination with the line of flight is reduced and the machine descends, the reverse action being obtained by pulling in. There is no operating mechanism employed for transverse control in the earlier Voisin machines, lateral stability being attained by the use of a number of keels which took the form of vertical partitions at regular intervals between the main planes, thus dividing the machine into a number of cells. This has recently been abandoned, however, in favor of the system of independent ailerons.

Power Plant. A 50- to 55-horse-power motor, placed on the rear of the central chassis and back of the main planes, drives a two-bladed metal propeller direct at a speed of 1,200 r. p. m., the propeller measuring 7.6 feet in diameter with a pitch of 4.6 feet. Several types of motors have been used.

General. The aviator's seat is placed on the central chassis in front of the motor and just back of the forward edge of the main planes. As a starting and landing chassis, two large pneumatic-tired wheels fitted with coil spring shock absorbers are fitted at the front and two at the rear. To avoid disastrous results, should the machine land at too sharp an angle, head-on, a small wheel is fitted to the front end of the chassis directly beneath the elevating rudder. The total weight is from 1,100 to 1,250 pounds, speed 35 m. p. h.; 23 pounds are lifted per horse-power, and 2.37 pounds per square foot of surface. The aspect ratio is 5.75 to 1. The use of the six vertical planes (two vertical walls of the rear cell and four vertical partitions between the two main supporting planes), Fig. 18, for steadying the machine transversely and keeping it to its course, are much lauded by Berget as superior to the Wright system of warping the planes, but experience appears to have proved to the contrary.

The Voisin type of biplane has recently been modified as follows: The vertical partitions have been done away with and ailerons are employed, together with a single plane, horizontal keel at the rear, instead of two planes. A 60-horse-power E. N. V. motor is employed, the total weight of the machine being about 1,170 pounds, giving a lift of 19.5 pounds per horse-power, and 3.27 pounds per square foot of surface. The aspect ratio is 5.13 to 1. This is a racing type of Voisin and is characterized by the elimination of most of the struts, cross wires, and other parts tending to increase the resistance to flight.

The regular Voisin biplane also has been altered by discarding the vertical partitions altogether, the design otherwise remaining the

Fig. 19. Voisin Biplane in which Paris-Bordeaux Flight was Made

same. This machine has a spread of 36.1 feet, a depth of 6.56 feet, a total area of 430 square feet, and a weight of 1,350 pounds. The motor employed is an eight-cylinder E. N. V. of 60 horse-power, carrying 22.5 pounds per horse-power and 3.14 pounds per square foot of surface. The aspect ratio is 5.5 to 1. In some of the more recent Voisin machines the front elevating rudder also has been discarded, Fig. 19.

Voisin Tractor Screw. This machine, Fig. 20, was first built in the latter part of 1909, and embodies several totally new departures in the construction of biplanes. It did not meet with particular

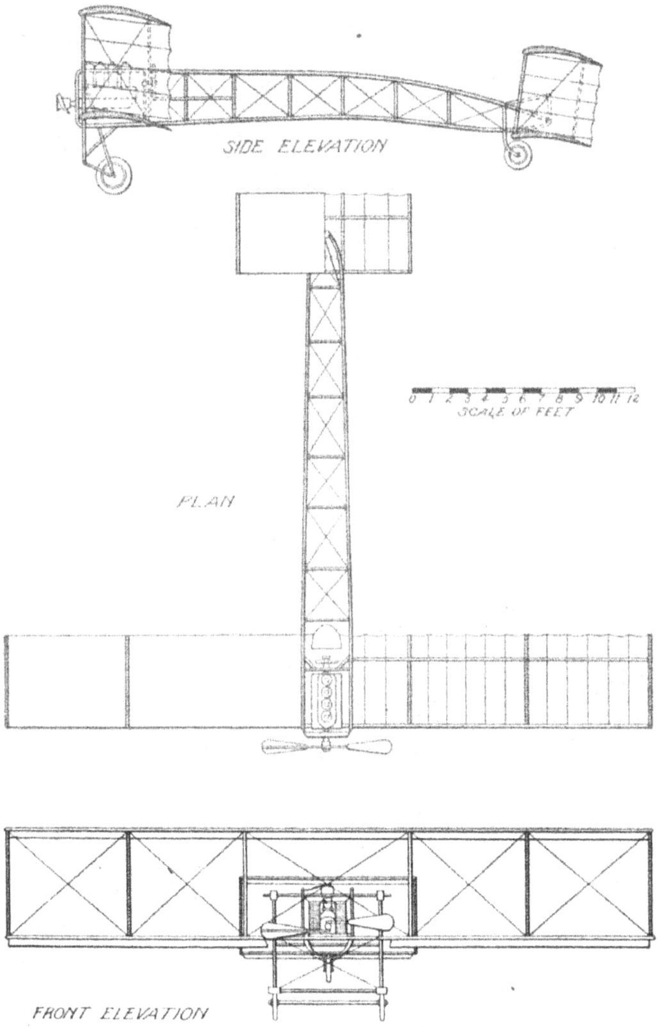

Fig. 20. Voisin Tractor Screw Type

success during 1910, although the Goupy and Breguet aeroplanes of the same type have been flown with great ease.

Frame. In this instance, the central chassis is extended a considerable distance to the rear, forming an "appendage." At the front are situated the motor and propeller, while directly behind the propeller is the main cell, with an auxiliary cell at the extreme rear. Ash, steel joints, and steel tubing are used throughout.

Supporting Planes. The supporting planes are two similar, directly-superposed surfaces with a spread of 37 feet, a depth of 5 feet, and an area of 370 square feet. By comparing the side elevations of the Voisin and Wright machines, the slight difference in the curvature of the planes, as well as their thickness, will be noted, though on comparing this feature in all of the machines illustrated, their striking similarity, as well as their close adherence to the pisciform contour of the plane—laid down by Colonel Renard as the most efficient shape for speed and stability—will be at once apparent.

Direction and Elevation Rudder. As these two elements are combined in the actual construction, they are accordingly described together. They are formed by the rear cell, consisting of two horizontal surfaces of about 80 square feet of area, and two vertical surfaces of about 50 square feet, the entire cell being pivoted on a universal joint so that it may be moved in any direction. The movement of the cell is controlled by cables leading to a large steering wheel in front of the aviator, the horizontal surfaces acting to elevate or depress the machine, and the vertical surfaces to change the direction of its travel. To ascend, the inclination of the cell relative to the line of flight is decreased; the leverage desired being the opposite of that necessary with a front elevation rudder. Four vertical partitions are placed between the main planes. There is no transverse control.

Power Plant. The power plant consists of a 40-horse-power, four-cylinder Voisin motor placed at the front end of the chassis and carrying directly on its crank shaft a two-bladed metal propeller 7.2 feet in diameter with a 4-foot pitch, which it drives at 1,300 r. p. m.

General. The chassis is mounted on two large pneumatic-tired wheels forward, fitted with shock-absorbing springs, and a smaller third wheel at the rear, while the aviator's seat is placed on the central frame at the rear of the main cell. The total weight is from

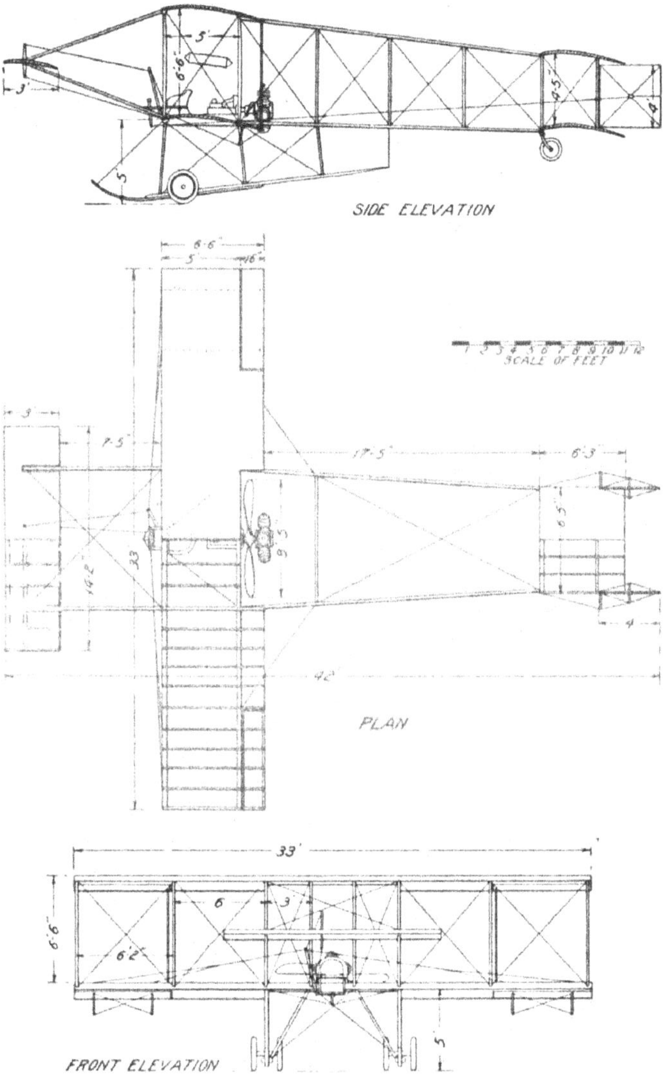

Fig. 21. Details of Farman III Biplane

800 to 950 pounds and the speed is said to be 50 miles an hour; 19 pounds per horse-power are lifted and 2.36 pounds carried per square foot of surface. The aspect ratio is 7.4 to 1.

Farman. The Farman machine, Fig. 21, has figured very prominently in the making of records and the winning of prizes, having been employed extensively by such aviators as Paulhan and White, as well as by Farman himself, Fig. 22. More than a hundred of the Farman biplanes had been built and put into use up to the end of 1910. It is a comparatively heavy type, and for a slow-moving, reliable machine it has proved very satisfactory.

Frame. The frame consists essentially of a main box-cell, somewhat similar in design to a Pratt truss, counterbalanced throughout with identical upper and lower chords, uprights of wood acting as compression members and cross wires as tension members, as is the case in all of the biplanes considered in this description of standard types. The supporting surfaces are analogous to the upper and lower decks of such a truss.

Supporting Planes. These supporting planes are practically identical with those of the machines already described, the surfaces themselves being made of continental cloth, stretched tightly over ash ribs. Their spread is 33 feet, depth 6.6 feet, and total area 430 square feet. The distance between the planes is 6.6 feet, which causes the machine to appear very much larger than the others by comparison and also gives it a very cumbrous look, the latter being accentuated by its very deliberate flight.

Elevation Rudder. The elevation rudder consists of a single, horizontal surface having an area of 43 square feet and is placed well out in front. It is hinged and braced to two sets of outriggers, firmly attached to the main cell, and is controlled by a large lever at the aviator's right hand. By pulling on this lever, the rudder is tilted upward and the machine rises, the method of control being almost instinctive and very easily acquired.

Direction Rudder. Two equal surfaces vertically placed, of an aggregate area of 30 feet, serve to control the travel of the machine. These surfaces move together and are operated by a pivoted lever on which the aviator rests his feet. By pressing so as to turn the lever to the left the machine alters its course in the same direction, the movement being transmitted to the rudder itself by cables.

Transverse Control. The control of the lateral equilibrium, *i. e.*, the tipping from side to side, is effected by the use of ailerons or "wing tips" consisting of four flaps constituting the rear ends of each plane. The operation of these wing tips is brought about simultaneously with that of the direction rudder through an arrangement identical with that on the Wright biplane, *i. e.*, a lever which may be moved in any direction, its forward and back motion actuating the rudder, while a sidewise movement operates the wing tips, from which it will be apparent that they are merely a modification of the Wright idea. This lever is connected by wires to the lower flap on each side and they are interconnected in the same manner with the flaps above them. When the machine is standing still the flaps merely hang loose and the wires relax, but when in flight the wind keeps them out and the wires are taut so that they may be controlled by the lever. The extra resistance these flaps or ailerons create is probably responsible in large measure for the decreased speed of the machine.

Keels. Two horizontal surfaces at the rear act as keels. Their combined area is about 80 feet, but as their angle of incidence is low the lift they exert is small, their only function being to steady the machine longitudinally.

Power Plant. The power plant consists of a 50-horse-power, seven-cylinder, air-cooled, rotary Gnome motor, mounted on a shaft at the rear of the lower plane. A two-bladed wood propeller of 8.5 feet in diameter by a 4.62-foot pitch is attached directly to it and revolves with the motor at a speed of 1,200 r. p. m.

General. The machine is mounted on two long skids forming part of the framework, similar to the Wright construction, and upon each of these skids is placed a pair of wheels. The latter are attached to rubber springs so that in starting the machine runs on them, but in alighting they give way, permitting it to slide on the skids. The total weight is from 1,100 to 1,350 pounds, the variation in this, as in every instance, being accounted for by the fact that it includes that of the aviator. The weight lifted per horse-power is 24 pounds, and 2.8 pounds per square foot of surface, while the speed is 37 miles per hour. The aspect ratio is 5 to 1.

New Models. In the foregoing, a description has been given of the original type of Farman biplane, numerous modifications

Fig. [illegible]

having been made in more recent machines, Fig. 22. The latter, for instance, are fitted with a single-surface direction rudder, instead of the twin surfaces mentioned. The elevation rudder is kept in front, but is made smaller, and in addition the rear end of the upper of the two fixed, horizontal keels at the rear is made movable conjointly with the front rudder to control the elevation of the machine. In some cases, only a single surface is used at the rear. One wheel has been substituted for the two formerly employed, the other characteristics of the machine remaining substantially as described.

The new racing Farman biplane is distinguished by the following features: The spread is reduced to 28 feet and the area to 350 square feet, while the total weight in flight is about 1,050 pounds. The lift is 21 pounds per horse-power, while that per square foot is the unusually high figure of 3 pounds. The aspect ratio is 4.2 to 1.

Another more recent type of Farman is the huge, new passenger-carrying machine which made the first four-passenger record. This has a spread of 47.6 feet and an area of approximately 540 square feet. The maximum total weight is nearly 1,750 pounds, or close to a ton, thus giving a capacity of 34 pounds per horse-power and a loading of 3.15 pounds per square foot of surface. The aspect ratio is 7.1 to 1.

In a still later type of the Farman, the ailerons are let into the wings and while they are hinged they are not permitted to hang down, as was formerly the case, this innovation being responsible for a decided reduction in the head resistance. Another type, brought out at the end of 1911, shows an entirely new form of stabilizing surfaces. These take the form of two pairs of long planes, one at each end of the main planes, and with their narrow edge to the wind, giving them a very small aspect ratio, though they have a comparatively large area. Each pair is held apart by struts and they are mounted on a vertical shaft, which is turned to swing them outward. The construction of the main cell in this machine does not exhibit any departures from the regulation Farman form, but in the machine with the set-in ailerons, which also made its debut at the Paris Salon at the end of 1911, the planes are "staggered," *i. e.*, the lower plane is very much shorter than the upper, and they are connected by diagonal steel struts, thus doing away with the maze of wire braces. A single surface tail is employed in connection with front and rear elevators and twin vertical rudders.

The Maurice Farman biplane differs somewhat from the machines just described (Henri Farman), the two brothers having at first operated independently. It is noteworthy for its remarkable duration performances. It was in one of the Maurice Farman biplanes that Tabuteau broke the 1909 world's duration record of 244 miles in 5 hours 3 minutes 5 seconds, by traveling 290 miles in 6:8:12 (October 28, 1910), which he increased on December 30, 1910, to 365 miles in 7:48:31, thereby winning the Michelin cup. The same machine also won the $20,000 prize for the flight from Paris with a passenger to the Puy de Dome, a mountain 4,800 feet high and 235 miles distant. Numerous attempts had been made to win this during three successive years. The Farman biplane covered the distance in 5:10:46, including a stop of 14 minutes, the time limit in which the prize could be won being six hours, which included circling the Arc de Triomphe in Paris and the steeples of the cathedral at Clermont-Ferrand near the finish as part of the conditions. The machine has a supporting surface of 635 square feet and an aspect ratio of 8 to 1. Its weight is 1,210 pounds and with a 60-horse-power Renault, eight-cylinder, air-cooled motor its speed is 48 miles per hour. The propeller is driven from the cam shaft instead of the crank shaft, so that at a motor speed of 1,800 r. p. m., it makes 900 r. p. m. Maurice Farman was the first to employ a covered body enclosing the seats and the engine, this construction now being considered essential for the comfort of the pilot and passenger on all Continental aeroplanes, though up to the beginning of 1912, it had not been made a feature of any of the American machines. The Farman control is very simple and effective. It consists of a hand wheel on a sliding shaft and a pair of pedals. Forward and backward motion of the wheel controls the angle of the elevator, while rotation of the wheel operates the rudders, the pedals actuating the ailerons. The wheel is vertical, its shaft passing horizontally through an automobile type of dash on which are mounted a clock, a gradient indicator, an aneroid barometer, and a recording barograph.

Sommer. In June, 1909, Roger Sommer purchased a biplane constructed by Henri Farman (the machine of Maurice Farman differs in design) and on July 3 he made his first flight. Scarcely a month later he held what was then the world's record for duration

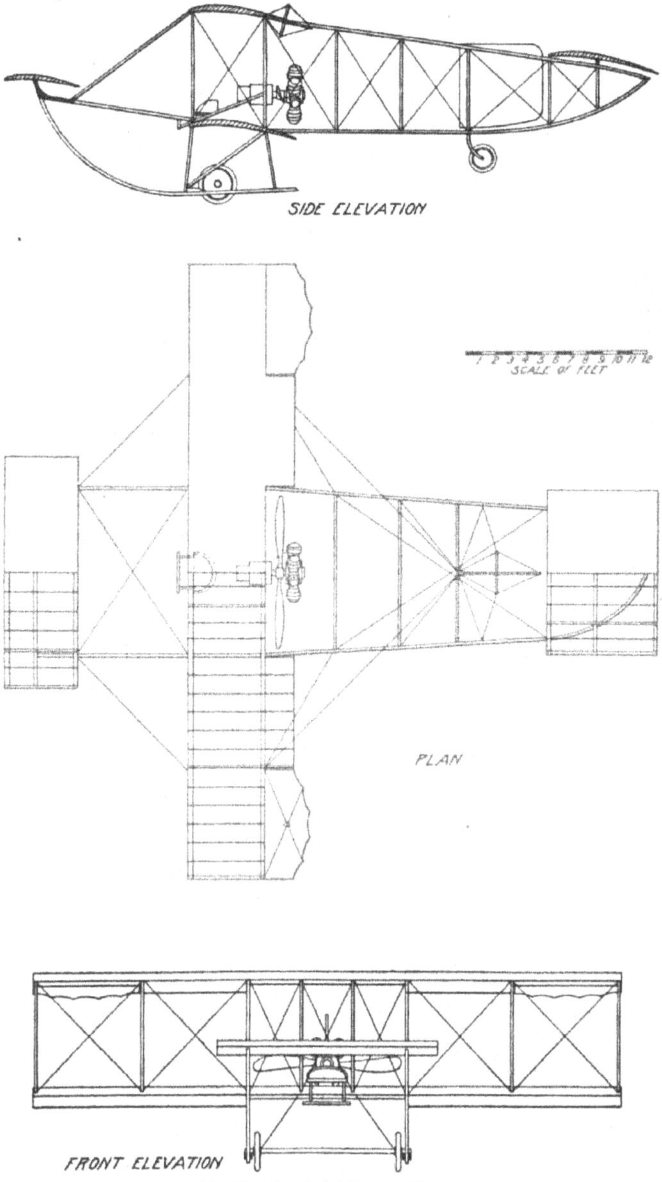

Fig. 23. Details of Sommer Biplane

of flight, having flown continuously for two and one half hours. His sudden jump into the ranks of the great aviators was remarkable and showed that, after all, it is not so hard to learn to fly well. He won many prizes at Rheims and Doncaster in 1909, but shortly afterward gave up flying on the Farman biplane and proceeded to design and build one of his own, Fig. 23. This was first tried out in January, 1910, and after a few days of experimenting he succeeded in making a long cross-country flight. The Sommer biplane is also operated by other prominent French aviators.

Frame. The construction of the frame is chiefly of hickory and ash with steel joints and steel tubing, its general character and appearance being similar to that of the Farman.

Supporting Planes. Two identical and directly-superposed rigid planes carry the machine, the surfaces being made of rubber cloth covering wood ribs. The sectional curvature of the surfaces is not so highly arched as in most other types, being more nearly, as in the Wright machine, a very even and gently sloping curve. The spread of the planes is 33 feet, their depth 5.2 feet, and their area 326 square feet.

Elevation Rudder. At a distance of 8.25 feet in front of the main cell, and supported on framing carried down to the skids, is placed the single-surface elevating rudder. This is governed by a large lever held in the aviator's left hand, which, when pushed out, turns down the rudder and, when pulled in, turns it up; thus, respectively, causing the aeroplane to mount or descend.

Direction Rudder. The direction rudder consists of a single surface of but 10 square feet in area, placed at the rear. It is operated by a pivoted foot lever similar to that of the Farman.

Transverse Control. Lateral equilibrium is secured by two wing tips, one placed at either end of the rear of the upper plane, as shown clearly in Fig. 24, there being no ailerons on the lower main plane as in the Farman. These are controlled by cables leading to a brace attached to the aviator's body. By leaning to the right, the wing tip on the left is pulled down, at the same time pulling up that on the right, causing the left end of the machine to rise and the right end to descend. Though not interconnected, the direction rudder and the transverse control are operated simultaneously by the operator, thus giving the same effect as is obtained in the Wright and

Farman machines by controlling these two elements from the same lever.

Keels. A single horizontal plane of 55 square feet area and of very light construction is placed at the rear and steadies the machine longitudinally. This plane is movable though it does not act as a rudder. A lever at the right hand of the operator, which automatically locks in place, enables the angle of incidence of this surface to be varied at will, thus increasing the attainable stability.

Power Plant. The power plant consists of the same type of rotary, air-cooled, seven-cylinder Gnome motor as employed on the Farman. It is placed at the rear of the main cell and is attached

Fig. 24. Sommer Biplane Equipped with Gnome Seven-Cylinder Motor

to a two-bladed wood propeller of 7.2 diameter by a 5.2-foot pitch, which it revolves at 1,200 r. p. m.

General. Two large wheels are attached forward and two small wheels at the rear of the chassis, the front wheels being held by rubber springs to two skids, built under the frame. The skids themselves are attached to the main frame by uprights, the joints being made of a springy sheet of metal bolted to the framing. This adds still further to the resilient character of the mounting. The seat for the aviator is placed on the front of the lower main plane at the center and is fitted more comfortably than on most other biplanes which had been built up to that time.

In more recent machines for racing purposes the two end panels of the lower surface of the Sommer have been eliminated, reducing the spread and cutting the area down to 256 square feet. The loading is 3.25 pounds per square foot.

Cody. Colonel Cody, an American, for a long time resident in England, is doubtless best known in this field through his connection with the successful operation of man-carrying kites several years ago. His work in this line for military scouting attracted considerable attention in England. In 1907, he commenced work on an aeroplane

Fig. 25. Cody Biplane Ready for Flight

of huge dimensions, Fig. 25. At first, the tests of this machine were very disappointing, but by his remarkable perseverance Colonel Cody turned failures into successes and finally, in the late summer of 1909, accomplished a superb flight of over an hour, establishing what was then the world's record for cross-country flight. The machine has been altered a number of times, and in its form as settled upon in the spring of 1910, Fig. 26, was the largest successful aeroplane in use.

Frame. Bamboo is employed extensively throughout the frame but all joints are wound with steel wire. In addition, there are a number of upright members of ash. At the center several members

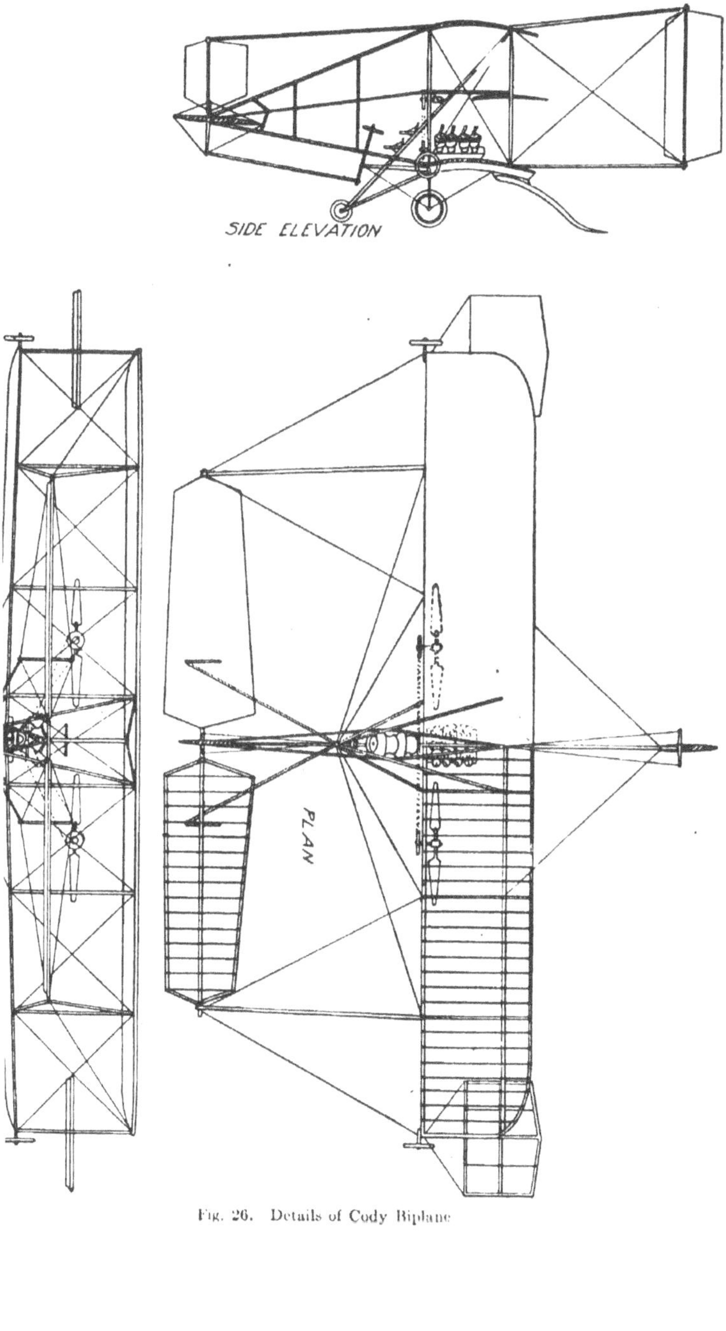

Fig. 26. Details of Cody Biplane

meet in the supporting chassis which is very heavily built. Steel wire is used for bracing.

Supporting Planes. The main planes are of rectangular form with rounded rear edges. They are identical and directly superposed, the surfaces being made of canvas tightly stretched over wood ribs. At the center, the distance between them is 9 feet, but they converge slightly toward either end where they are separated by only 8 feet. The spread is 52 feet, the depth 7.5 feet, and the area 780 square feet.

Elevation Rudder. At the front of the machine, supported by large bamboo outriggers from the central cell, are two equal surfaces placed on either side of the center. They are jointly movable and serve to control the elevation of the machine. They are governed by the forward or backward movement of the stanchion to which the steering wheel is attached, in the same manner as on the Curtiss. If the aviator wishes to rise, he pulls the wheel toward him. This motion, by means of a lever system, causes the elevating rudder surfaces to be tilted upward to the line of flight and the machine ascends.

Direction Rudders. Two direction rudders are employed, a large one at the rear and a small one in front, the former constituting the main rudder. These rudders are moved together by a steering wheel and cables as in a motorboat. Their combined area is about 40 square feet.

Transverse Control. Two balancing planes of 30 square feet area, one placed at either end of the main cell, control the transverse inclination of the machine. They are moved inversely by cables leading from the steering gear and operate in the same manner as the ailerons of the Curtiss machine and the wing tips of the Farman and Sommer biplanes, one balancing plane being turned up while the other is turned down. Lateral stability is also controlled by the inverse movement of the two halves of the elevation rudder, the one on the depressed side being elevated while the other is turned down. There are no keels on the Cody biplane, all surfaces serving either to lift or direct the machine.

Power Plant. The power plant is an eight-cylinder, 80-horsepower E. N. V. motor placed near the forward edge of the lower main plane and directly back of the aviator. It drives two two-bladed wood propellers mounted on shafts located at their front end half way between the main planes. These are driven in opposite directions by means of chains, as in the Wright biplane. These propellers have a diameter of 8.25 feet and a pitch of 6 feet; and are revolved at 600 r. p. m.

General. The mounting consists of a large pair of wheels which carry most of the weight, a small wheel in front and a skid at the rear. Wheels are also attached to the outer ends of the lower plane to carry the machine easily over the ground should it alight on end. The total weight is from 1,900 to 2,100 pounds; speed 37 miles per hour; 25 pounds per horse-power are lifted and 2.57 pounds per square foot of supporting surface. The aspect ratio is high—7 to 1. Seats are provided for the aviator and for one passenger, both being placed low at the center of the front of the main cell, that for the passenger being higher than that for the aviator, as it is designed for the use of an observer in war time.

Since the machine was first built, the E. N. V. motor has been replaced by two 50-horse-power, four-cylinder Green motors, both driving a single propeller instead of the twin propellers formerly used. Either motor can be operated independently, the advantage of this arrangement being that if one motor breaks down while in flight the other can still be used to drive the machine.

MONOPLANES

Antoinette. Up to the time of the present writing, the Antoinette, Fig. 27, is the largest monoplane in use and its construction is distinguished by a number of features not found on others. Levavasseur, designer of the Antoinette motor and motorboats, is credited with the design of this type. After building some experimental machines, notably the Gastambide-Mengin monoplane, the Antoinette IV was built for Hubert Latham. This machine was at first controlled transversely by means of wing tips, but the warpable surface, or Wright control, has since been adopted. The Antoinette is remarkably well built from an engineering standpoint and has been successfully operated by M. Latham in high winds, though not as strong as the gale in which the two Wright biplanes were blown backward 30 and 40 miles from Belmont Park at the International Meet, despite all they could do. The Antoinette is also flown by other prominent French aviators and several of the machines have been purchased by the French War Department.

Frame. The frame is of long, narrow, girder-like construction, Fig. 28, of cedar, ash, and aluminum, carrying at its forward part the main plane, the "nacelle" or car for the aviator, and at its extreme front end the propeller, while at the opposite end are placed the rudders, the longitudinal dimensions of the machine being in excess of 36 feet, or almost three fourths as much as its spread. The arrangement of the planes and rudder, as well as the location of the motor, is similar to that in all the monoplanes described here with the exception of the Pfitzner.

Supporting Plane. The supporting plane consists of a single surface divided in half, the two sections being of trapezoidal shape, placed at a slight dihedral angle to each other. They are constructed

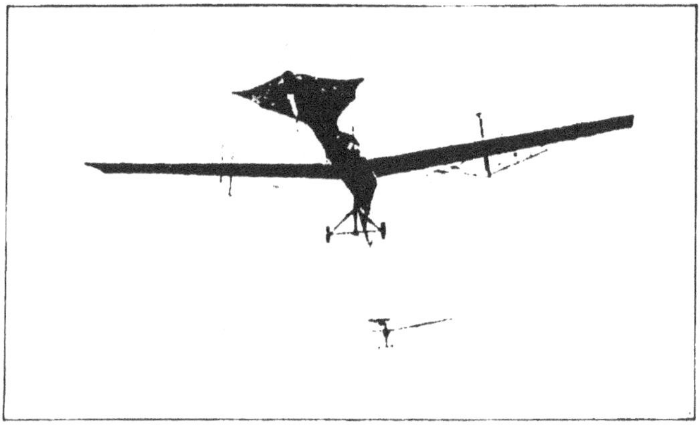

Fig. 27. Two Antoinette Monoplanes Competing at Belmont Park, 1910

of rigid trussing, nearly a foot thick at the center and covered over and under with a smooth, finely-pumiced silk. The plane is also braced from a central mast. The spread is 46 feet, the average depth 8.2 feet, and the surface area 370 square feet.

Direction Rudder. The direction rudder consists of two vertical triangular surfaces at the rear and measures 10 square feet in area. These surfaces are moved jointly by means of wiring cables worked by a lever operated by the aviator's feet. When this lever, which moves in a horizontal plane, is turned to the left, the machine will change its course in the same direction.

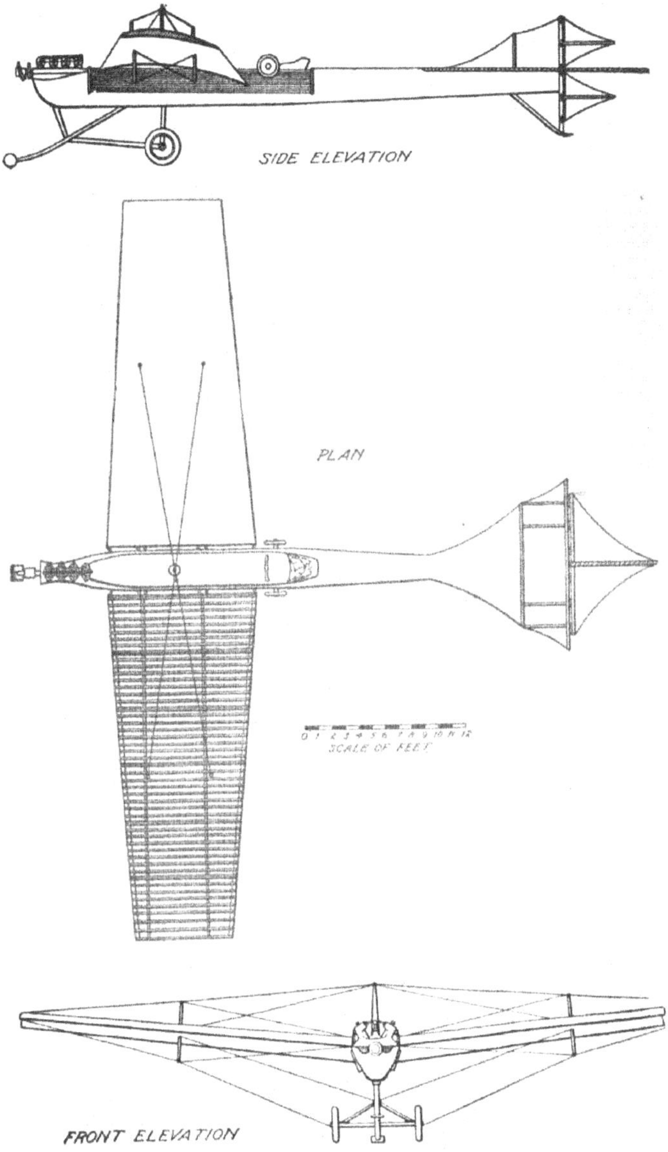

Fig. 28. Details of Antoinette Monoplane

Elevation Rudder. The elevation rudder has an area of 20 square feet and is also triangular. It is placed at the extreme rear in order to provide the maximum leverage, and is controlled by cables leading round a drum attached to a wheel at the aviator's right hand. To ascend, the wheel is turned up. This causes the inclination of the elevating rudder with regard to the line of flight to decrease, and the machine, therefore, rises.

Transverse Control. Lateral stability is maintained by warping the outer ends of the main plane in much the same manner as in the Wright machine, except that the front ends of the plane are movable and the rear ends are rigid throughout in the Antoinette, the reverse being the case in the Wright. Through cables and a sprocket placed at the lower end of the central mast, the warping is controlled by a wheel at the aviator's left hand. To correct a downward inclination at the right, the right end of the wing is turned up and at the same time the left end is turned down, restoring the balance.

Keels. At the rear, leading up to the rudders, are tapered keels, both horizontal and vertical, that add greatly to the bird-like appearance of the Antoinette in flight.

Power Plant. The power plant is an eight-cylinder, V-type four-cycle, water-cooled Antoinette motor of 50 horse-power, the radiator taking the form of two banks of tubes placed along either side of the car. The motor carries on the forward end of its crank shaft a two-bladed, metal propeller, 7.25 feet in diameter by 4.3 feet pitch, which it drives at 1,100 r. p. m.

General. The chassis is mounted on a pair of pneumatic-tired wheels attached to the central mast by a pneumatic spring. In addition, a single skid is placed forward to protect the propeller in landing, and another at the rear. The seat for the aviator is placed in the frame back of the main plane and about 8 feet directly behind the motor, a seat for a passenger being provided in front of and slightly lower than that for the aviator. The sides of the space are walled with canvas, affording the aviator and passenger more protection than is usually provided. The total weight is 1,040 to 1,120 pounds, the speed 43 miles per hour. Thirty pounds are lifted per horse-power and 3.96 pounds per square foot of supporting surface. The aspect ratio is 6 to 1.

In a later machine, the spread is 49.3 feet, the area 405 square feet, and the total weight 1,200 to 1,350 pounds, 27 pounds being lifted per horse-power, and 3.33 pounds per square foot of surface. The aspect ratio is 6 to 1. A new 100-horse-power type is also employed for racing, this machine being fitted with the Antoinette sixteen-cylinder, V-type motor. The newer models of the Antoinette differ so radically that they have been described in the article devoted to special types.

Santos-Dumont. The first sustained flight of a motor-driven aeroplane in Europe was made by M. Santos-Dumont on November

Fig. 29. Santos-Dumont's Earliest Aeroplane with Which He Made the First Power Flight in Europe

12, 1906, in a biplane of his own design, Fig. 29. In 1907 he began work on a monoplane and after a great deal of experimenting succeeded in evolving the Demoiselle, Fig. 30, so-called owing to its diminutive size, as it is the smallest aeroplane in use up to the present writing. It is extremely simple and compact and many of them are flown abroad. Some of Santos-Dumont's earlier attempts were based on principles attractive in theory, but which experience has shown to be erroneous. Chief among these are the use of a sharp dihedral angle for the supporting surfaces and a very low center of gravity to simulate a pendulum. As shown by the Wright Brothers' experiments, while a pendulum may give a certain stability in a state of perfect rest or when flying straightaway in a dead calm, it exaggerates oscillation, once the latter is set up, and is entirely destructive of stability. Planes set at a dihedral angle give neither the same lifting power nor an amount of stability equal to a surface of the same dimensions that is made perfectly flat laterally. This is the case in all the biplanes described here and some of the monoplanes, the supporting surfaces of the Bleriot and Antoinette being set at a slight dihedral angle, however. This characteristic is still

Fig. 30. Santos-Dumont Demoiselle, the Smallest Man-Carrying Aeroplane

strongly marked in the Santos-Dumont monoplane, but the motor has been placed on a level with the supporting surfaces. The lack of stability of this machine was very marked as compared with both the biplanes and monoplanes taking part in the International Meet near New York, both its pitching and rocking reaching extreme angles and continuing throughout the flight. When compared with the larger machines in the air, it appeared almost like a sparrow among eagles, and the difference in the character of their action in flight was also similar. At no time did Audemars or Garros leave the ground more than 30 or 40 feet below, when flying the Demoiselle monoplanes on the occasion in question.

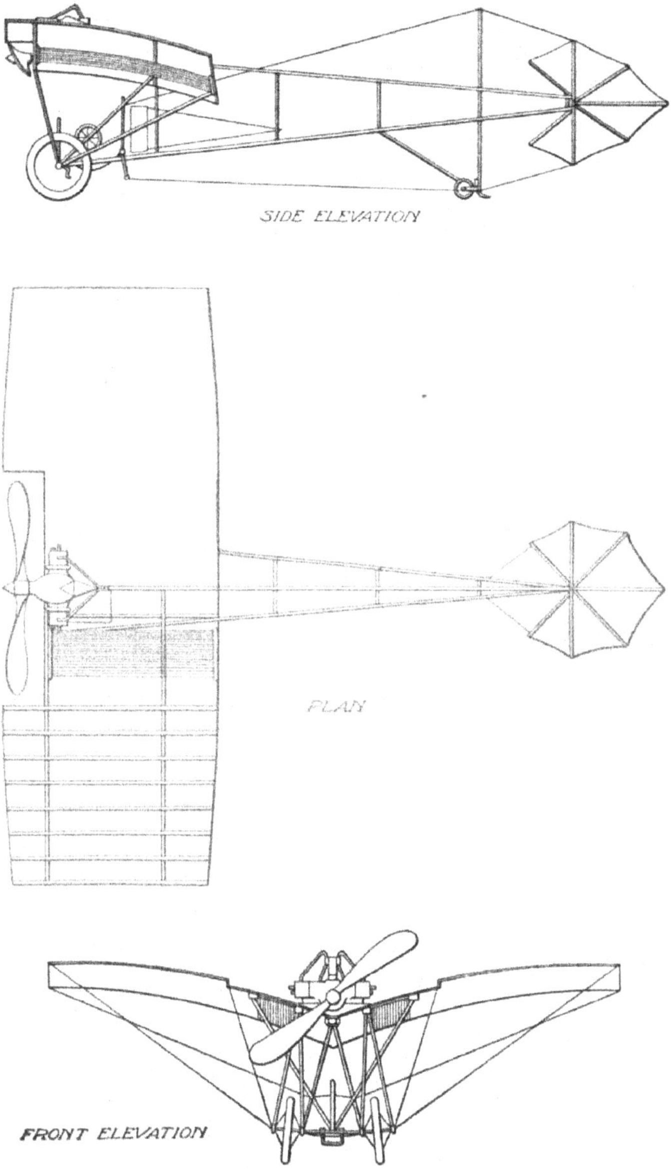

Fig. 31. Details of Santos-Dumont Demoiselle

Frame. The frame is triangular in form, Fig. 31, with its apex at the rear and is composed of bamboo with steel joints and several members of steel tubing.

Supporting Planes. Owing to the curvature of the supporting surfaces closely approximating the arc of a circle, there are really two planes joined at their inner ends. They consist of a double layer of silk tightly stretched over bamboo ribs, the whole being braced by steel wires led to the central frame. The spread is 18 feet, the depth 6.56 feet, and the area 113 square feet.

Direction and Elevation Rudders. The direction and elevation rudders are combined at the rear in the form of two fan-shaped surfaces, one vertical and the other horizontal, swung on a universal joint at the point of the triangular frame. The elevating rudder has an area of 21 square feet, while the direction rudder is somewhat smaller. A lever at the aviator's right hand controls the elevating rudder, while a wheel at the left operates the direction rudder. To ascend, the tail is moved up and to the right, to alter the line of travel in that direction. There are no keels.

Transverse Control. Transverse control is accomplished by warping the main planes, their operation being governed by a lever at the back of the aviator which fits into a pocket sewed into his coat. If the machine should suddenly tip up on the left, the aviator, by moving quickly in that direction, could pull down the plane on the right and increase the angle of incidence on that side. It will be seen from the foregoing that in flight he is kept pretty busy. The flexibility of the ribs of the planes permits them to warp without any special constructional details for that purpose.

Power Plant. A 30-horse-power, two-cylinder, horizontal-opposed water-cooled Darracq motor drives a two-bladed Chauviere wood propeller 6.9 feet in diameter by 6-foot pitch at 1,400 r. p. m., although Clement-Bayard and Panhard motors are also used on this machine.

General. The machine is mounted on two rigidly attached pneumatic-tired wheels at the front and a single small skid at the rear, the aviator's seat consisting of a strip of canvas placed across the frame and located directly beneath the motor. The propeller, instead of extending forward beyond the main planes, revolves in a rectangular opening cut in the latter. The total weight is from 330 to 370 pounds, speed 52 miles per hour. Twelve pounds are lifted

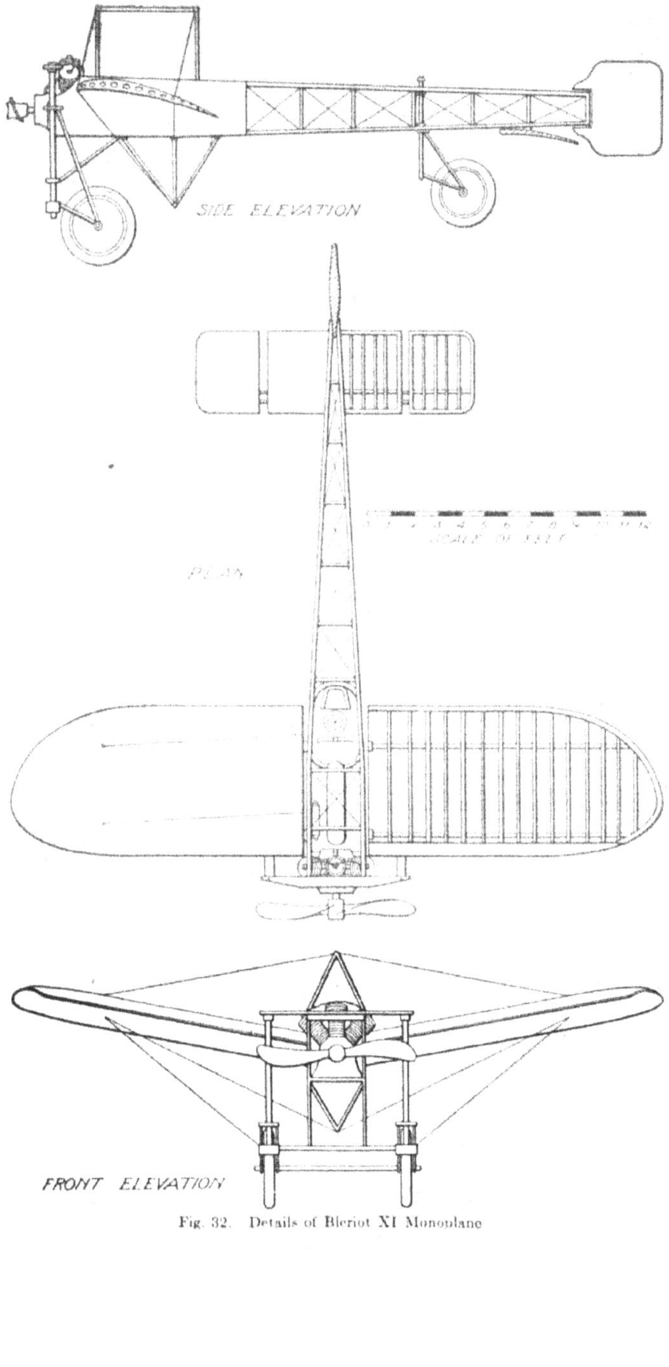

Fig. 32. Details of Bleriot XI Monoplane

per horse-power and 3.1 pounds per square foot of surface. The aspect ratio is 2.7 to 1.

Bleriot XI. In 1906, M. Louis Bleriot constructed and flew the first successful monoplane built. The two years following were devoted to experimental work, during which period a number of various modifications of the original were built until, in 1908, Bleriot succeeded in making a number of extended flights in his large monoplane, No. 8 Bis. In July, 1910, he made his sensational cross-channel trip, starting from Calais and landing near Dover. This flight was accomplished in the No. XI type machine, Fig. 32, a small one-passenger monoplane which is very simple and has come into widespread use abroad. Delagrange, Le Blanc, De Lesseps, Le Blon, Balsan, and Guyot are among some of the noted French aviators who have flown Bleriot monoplanes, two of whom have been killed in their operation. More than one hundred and forty of these machines were manufactured and sold during the year ending with August, 1910.

Frame. The frame consists essentially of a long central body of tapering construction to which the planes and rudder are attached. The framework is very lightly but strongly built of wood and is cross braced with steel wires throughout.

Supporting Plane. The main plane is placed at the forward part of the central frame and is divided in half, each section being mounted on either side of the central frame by socket joints. The halves are thus readily detachable at that point and when not in use are dismounted and placed in a vertical position along the frame so as to make the machine as a whole occupy very little room. The surfaces consist of wood ribs covered both above and below by Continental rubber fabric. The curvature is more pronounced than in most other types, with the exception of the Demoiselle, and a sharp front edge is obtained by the use of aluminum sheathing at that point. The two halves of the main plane are set at a slight dihedral angle. Their spread is 28.2 feet, depth 6.5 feet, and surface area 151 square feet. They are braced both above and below by steel wires led to the central frame.

Direction Rudder. The direction rudder consists of a very small plane having only 4.5 square feet of area and is placed at the extreme rear. It is controlled by a foot in the manner already described in some of the foregoing machines.

Elevation Rudder. The elevation rudder is divided into two parts, one half being mounted at each extremity of a fixed horizontal keel. It has 16 square feet of surface and is operated by the longitudinal movement of a bell crank device. This takes the form of a universally-pivoted lever placed in front of the operator, and is normally vertical. At the lower extremity of the lever is fixed a dome or hood-shaped piece of metal to which the wires are attached, at the same time protecting them from entanglement in the aviator's feet. To ascend the aviator pulls the lever toward him, and to descend pushes it from him.

Transverse Control. Lateral equilibrium is maintained by warping the main planes, the structure of the latter enabling them to be twisted as in the Wright machine, though in this case they warp about the bases which are rigidly attached to the main frame by the socket joints mentioned. The two halves are warped inversely by the side to side motion of the bell crank, *i. e.*, if the machine should tip up on the right, then the bell crank would be moved to the right. This would increase the incidence of the lowered side and at the same time decrease that of the raised side, thus righting the machine. The combination of this side to side movement of the bell crank with the movement of the foot lever controlling the direction rudder is used in turning.

Keels. To preserve the longitudinal stability, a single, fixed, horizontal keel is placed at the rear. Its area is 17 square feet.

Power Plant. The power plant is a three-cylinder, fan-shaped Anzani motor, developing 23 horse-power. It is of the air-cooled type and is placed at the forward end of the central frame. It drives a two-bladed wood propeller of 6.87 feet in diameter by 2.7-foot pitch direct at 1,350 r. p. m. Most of the more recent Bleriot monoplanes have been fitted with 50-horse-power Gnome, seven-cylinder, rotary, air-cooled motor.

General. The machine is mounted on an elastic chassis with two large rubber-tired wheels forward and a small wheel rear. The springs are made of thick rubber rope, affording great elasticity and strength with small weight. The aviator's seat is back of the main plane.

The total weight is from 650 to 720 pounds and the speed is 36 miles per hour with the Anzani motor and 48 miles per hour with the

Gnome motor; 29 pounds are lifted per horse-power and 4.5 pounds per square foot of surface, this ratio being unusually high. The aspect ratio is 4.35 to 1.

Later Types. In the later Bleriot machines, the elevating rudder is of different form, being attached at the rear edge of a tapering keel much larger than that formerly used. The small wheel at the rear has been replaced by a skid and the overall length of the central frame has been shortened considerably. The regular one-passenger type of this monoplane has further been altered to the new No. XI Bis, in which the sectional curvature of the planes has been made very nearly flat on the under side. This change has been found to

Fig. 33. Bleriot Two-Passenger Monoplane

greatly decrease the dynamic resistance of the machine without seriously impairing its lift. There are two new models of this machine which have been very successful. They are the No. XI 2 Bis, a two- or three-passenger machine, Fig. 33, and the No. XI racing model, Figs. 34 and 35. The former has a spread of 36 feet, a depth of 7.6 feet, a surface of 270 square feet, and a weight in flight of about 990 pounds. In other respects it resembles the No. XI Bis. 19.8 pounds are carried per horse-power and 3.68 pounds per square foot of surface. The aspect ratio is 4.75 to 1. The *type de course*, or No. XI racing model, is the machine on which Morane established the record of almost 69 miles per hour. It has a very short body, flat planes, and a reinforced frame. The surface has been reduced to 129 square feet and it is equipped with one of the new 100-horse-power,

Fig. 34. Bleriot Racing Model in Fl'ght

fourteen-cylinder Gnome, rotary, air-cooled motors. The total weight is about 750 pounds; only 7.5 pounds are carried per horse-power and as much as 5.76 pounds are lifted per square foot of surface.

Fig. 35. Bleriot Rounding a Pylon in International Race for Gordon-Bennett Cup

Bleriot XII. The Bleriot XII is a passenger-carrying type which differs in construction from those just described. With one of these large machines, M. Bleriot made the first flight in an aeroplane carrying three passengers. It has since come into general use, more than thirty of them having been built.

Frame. The long central frame of wood, Fig. 36, braced in every panel by steel cross wires, is very deep forward and tapers gracefully to a point at the rear.

Supporting Plane. On the upper deck of the central frame at the front is placed the main plane which is continuous and perfectly horizontal. Its structure is similar to that of the No. XI and it is braced by a number of wires from the frame. The spread is 30.2 feet, the depth 7.6 feet, and the total area 228 square feet.

Direction Rudder. A single surface placed at the rear extremity of the vertical keel is used for this purpose. Its area is only 9 square feet and it is operated in the same manner as on the No. XI.

Elevation Rudder. The elevation rudder also consists of a single surface of 20 square feet area and placed at the extreme rear. It is operated by the movement of a bell crank, as already described.

Transverse Control. The main surfaces are warped inversely, exactly as in the No. XI, a small surface under the aviator's seat also assisting in the lateral balancing. A horizontal keel of 21 square feet area is placed on the framework at the rear, but somewhat forward of the elevating rudder.

Power Plant. The power plant consists of a 60-horse-power, eight-cylinder E. N. V., air-cooled motor, placed on the frame under the main plane. By means of a chain transmission it drives an 8.8-foot propeller mounted on a shaft at the edge of the main plane. The propeller has an unusually long pitch—9 feet—and turns at only 600 r. p. m.

General. The mounting is similar to that of No. XI, while the seat or bench for three persons is placed under the main plane and back of the motor. The total weight is from 1,150 to 1,300 pounds; speed 48 miles per hour; 21 pounds are lifted per horsepower and 5.3 pounds per square foot of surface. The aspect ratio is 4 to 1. Bleriot is one of the most prolific designers of monoplanes, and it would require a volume to describe them. The Bleriot Limousine or "aerial taxi" is described under special types.

Grade. Herr Grade has the distinction of being one of the first German aviators to design and build an aeroplane. In the fall of

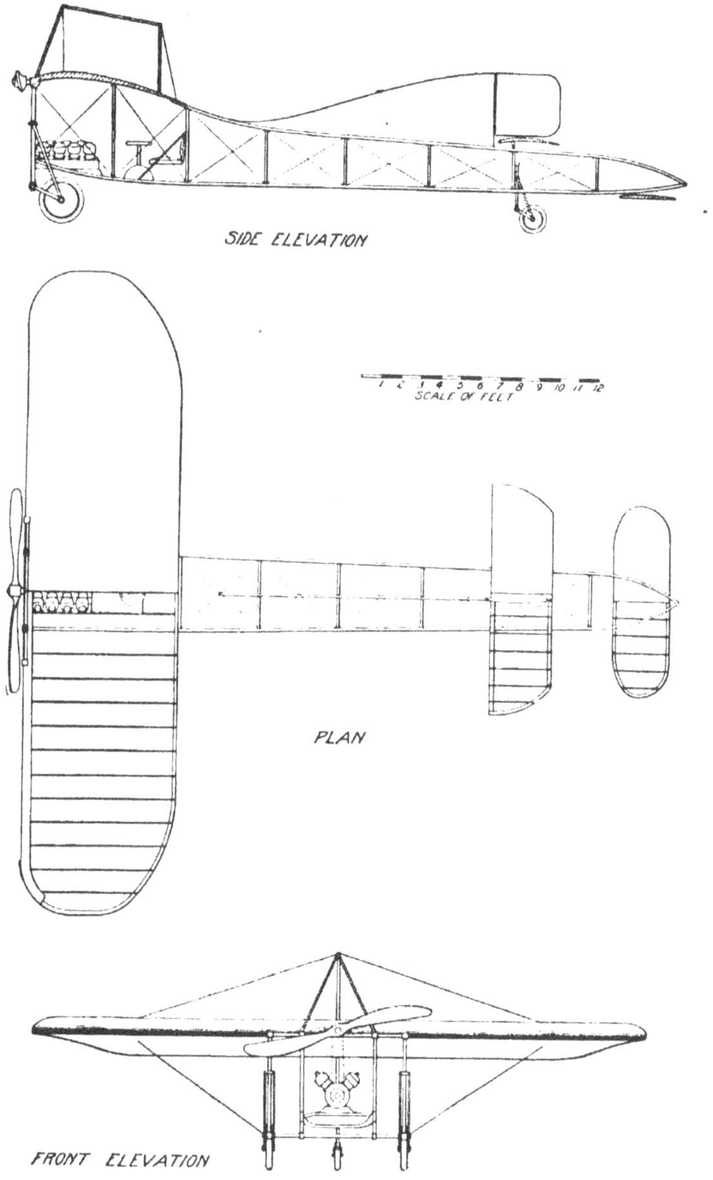

Fig. 36. Details of Bleriot XII Monoplane

1909, he began flights on his interesting monoplane, Fig. 37, and on October 30, 1909, won the Lanz $10,000 prize for a German-built

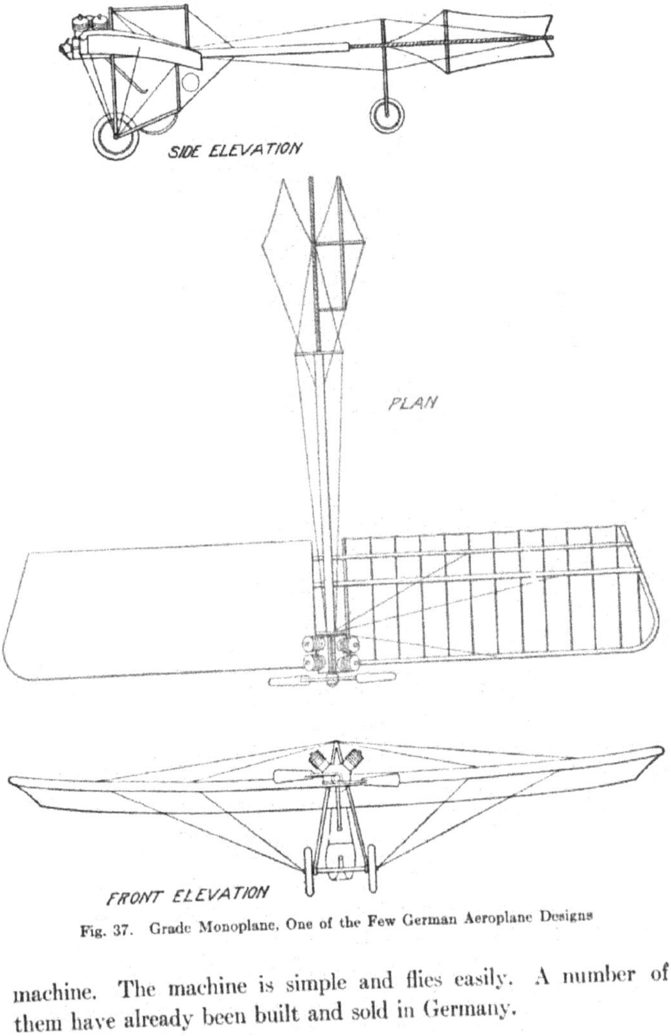

Fig. 37. Grade Monoplane, One of the Few German Aeroplane Designs

machine. The machine is simple and flies easily. A number of them have already been built and sold in Germany.

Frame. The frame is remarkable for the simplicity of its construction, consisting of a main metal tube chassis at the front from which a long thick member supporting the rudders is run out to the rear.

Supporting Plane. The main surface is made of Metzler rubber fabric stretched over a bamboo frame. The surface is very flexible and the two ends are turned up slightly from the center. The curvature is almost the arc of a circle and the section is very thin. The spread is 30 feet, depth 7 feet, and area 208 square feet.

Direction Rudder. The direction rudder consists of a single, flexible surface of 16 square feet area, carried at the rear and controlled by a lever. The surface itself is not hinged, but is bent in the direction desired by the lever and wire connections.

Elevation Rudder. The elevation rudder also consists of a single surface placed at the rear. It has an area of about 20 square feet and like the direction rudder its operation depends upon its flexibility. It is controlled by a large lever universally pivoted on the frame above the aviator. To rise, this lever is pulled up, and to descend, it is pushed down, thus bending up or down the rear horizontal surface.

Transverse Control. Warping the main planes is resorted to, the operation being similar to the Bleriot, which, in turn, is patterned after the Wright.

Keels. The tapering ends of both the direction and elevating rudders can be considered as keels, an additional vertical keel being placed forward, both above and below the main plane.

Power Plant. A four-cylinder, V-type, air-cooled motor of 24 horse-power is placed at the front edge of the plane. It drives direct at 1,000 r. p. m. a two-bladed metal propeller 6 feet in diameter by a 4-foot pitch.

General. Two wheels are employed forward and one rear for the mounting, no springs being provided. The front wheels are provided with a brake to bring the machine to a quick standstill after alighting, this being an important feature where the space is limited. The seat is placed under the main plane and consists of a hammock-like piece of cloth which is very light and very comfortable. The total weight is from 350 to 450 pounds and the speed approximately 44 miles per hour; 17 pounds are lifted per horse-power, and 1.9 pounds per square foot of surface. The aspect ratio is 3.2 to 1.

Pelterie. By many, the Pelterie monoplane, Fig. 38, is considered to be one of the most perfect types of aeroplane. Great

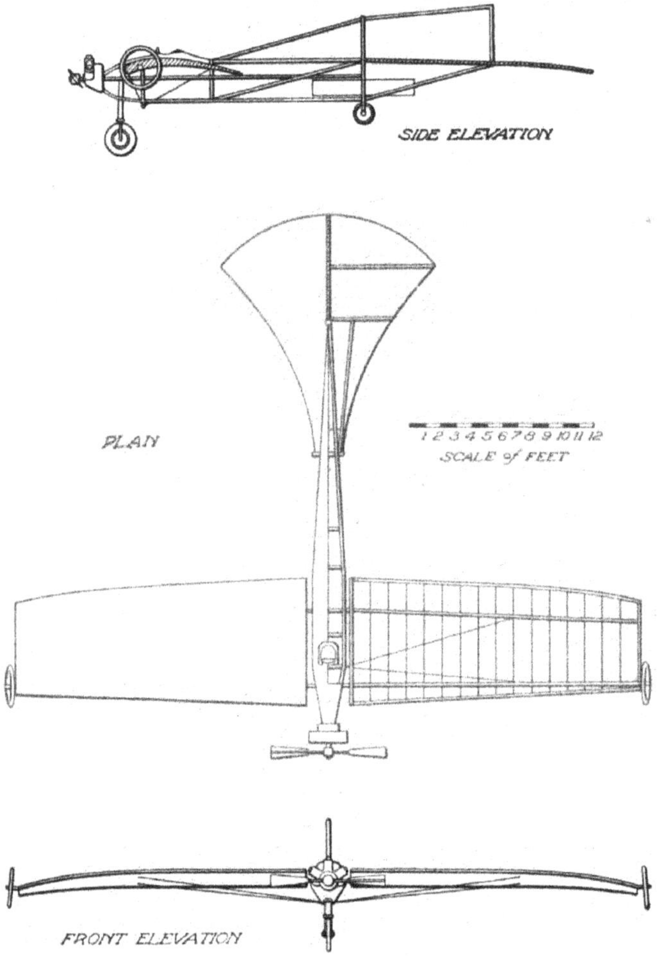

Fig. 38. Pelterie Monoplane

care is shown in its construction and finish, but owing to motor troubles, it has never flown for any length of time. Its designer,

Robert Esnault Pelterie, is one of the foremost French aviation scientists, and previous to building this machine, he conducted a lengthy series of gliding experiments of considerable interest.

Frame. The central frame, somewhat similar in form to a bird's body, is made largely of steel tubing and is quite short. All exposed parts are covered with Continental cloth.

Supporting Plane. The main supporting surface is particularly strong and solid, being made of steel tubing carrying wood ribs covered with Continental cloth. The curvature is very similar to that of a bird's wing, and transversely the surface curves downward, dihedrally from the center. Very little bracing is necessary. The spread is 35 feet, depth 6.1 feet, and the area 214 square feet.

Direction Rudder. The direction rudder consists of a vertical rectangular surface of 8 square feet area placed below the central frame at the rear. It is operated by a lever at the aviator's right.

Elevation Control. There is no elevation rudder in the Pelterie monoplane, the elevation of the machine being accomplished by changing the angle of incidence of the main planes themselves. To ascend, for instance, the aviator pulls the lever in his left hand toward him. This increases the angle of incidence of the plane and accordingly increases the lift, causing the machine to rise.

Transverse Control. Each half of the main plane is warpable about its base, transverse equilibrium being maintained by the inverse warping of the planes in the usual manner. In turning, both the left-hand lever controlling the warping planes and the right-hand lever controlling the direction rudder are simultaneously moved to the side desired. It is worthy of note here, that of all aeroplanes employing the Wright system for maintaining lateral stability—and there are very few that do not—none of them combines the control in one lever in the same ingenious manner as found in the Wright machine.

Keels. Vertical and horizontal keels consisting of gradually tapering surfaces are fixed to the frame and aid in preserving stability. The rear horizontal keel, shaped like a bird's tail, has an area of 20 square feet.

Power Plant. The power plant consists of a seven-cylinder, fan-shaped, air-cooled R. E. P. (Robert Esnault Pelterie) motor of very ingenious design. It is placed at the front and drives direct a four-bladed aluminum and steel propeller at 900 r. p. m. Its diameter is 6.6 feet and its pitch 5 feet.

General. The mounting consists of a large single wheel carried on a combined hydraulic and pneumatic spring at the center of the front, with a smaller wheel on the same center line at the rear. Wheels are also placed at the outer ends, or tips, of the supporting planes, so that when first starting to run along the ground, the machine is inclined. The seat is placed in the frame, and protected on all sides, the aviator's shoulders coming flush with the supporting surfaces. The total weight is from 900 to 970 pounds; speed 39 miles per hour; 27 pounds are lifted per horse-power and 4.4 pounds carried per square foot of surface. The aspect ratio is 5.75 to 1.

In a later model of the R. E. P. the fuselage is entirely of steel tubing connected by welded joints and the whole strongly trussed. Each wing is composed of two ash spars covered by red Continental rubberized fabric. The method of attaching the wings to the fuselage is a distinctive feature. In most monoplanes the ends of the spars are let into the fuselage, but in this case they are attached to the body by means of joints. This prevents the portions of the wings near the fuselage from having to endure abnormal stresses due to their attachment in case the supporting stays should become slack. This arrangement also permits the dihedral angle between the wings to be varied slightly. The lower stays of the rear spar are attached to an oscillating lever mounted on ball bearings and controlled by the wing-warping lever, while the lower stays attached to the front spar and supporting the wings in flight are steel cables covered with cloth. The tail fins, elevator, and rudder are demountable, being composed simply of steel tubing covered with fabric. The well-developed horizontal tail fins, being distant from the center of gravity, give great longitudinal stability to the machine. The elevator, which forms the prolongation of the horizontal empennage or tail, is divided into two parts by the rudder, forward of which is the vertical keel.

Pfitzner. The Pfitzner machine, Fig. 39, represents a radical departure from all other aeroplanes in some of its features, while it differs from other monoplanes in the placing of the aviator, motor, and rudders. It was built in the early part of January, 1910, by A. L. Pfitzner at the Curtiss factory in Hammondsport, New York.

It was the first to employ the comparatively simple and efficient method of transverse control by means of sliding surfaces, and while

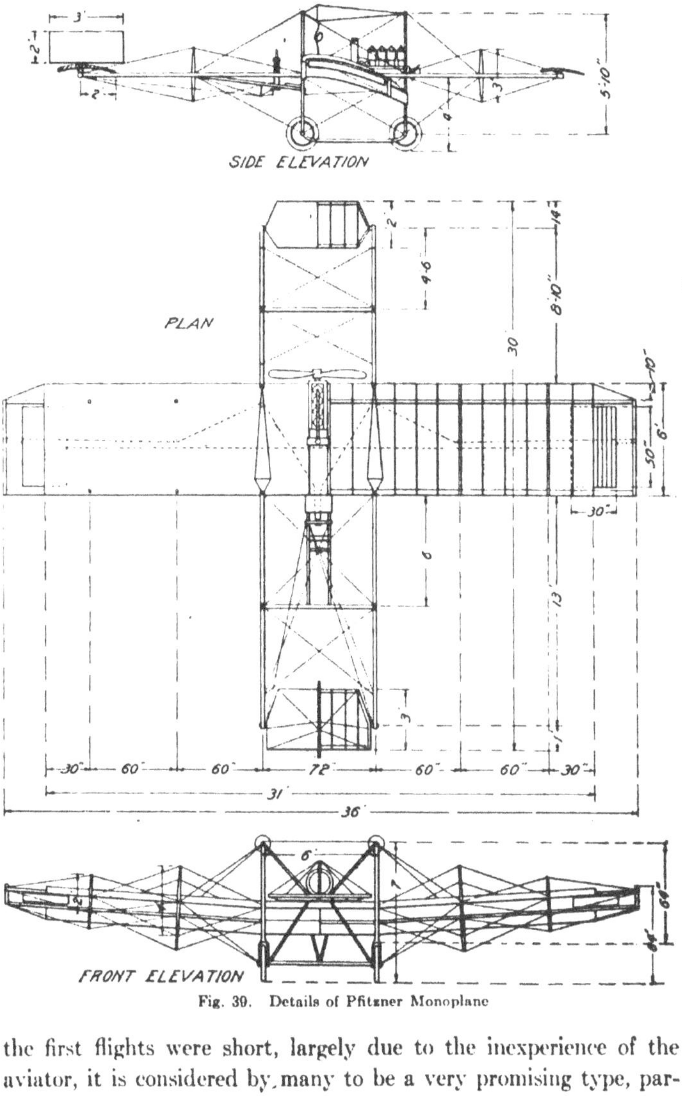

Fig. 39. Details of Pfitzner Monoplane

the first flights were short, largely due to the inexperience of the aviator, it is considered by many to be a very promising type, particularly as it does not conflict with the Wright system in any way.

Frame. The framework is largely a combination of numerous king-post trusses with spruce compression members and wire tension members. The framework is open throughout, enabling quick inspection and easy repairs. At the center, the chassis is mainly composed of steel tubing.

Supporting Plane. The main supporting plane, a 5-degree dihedral angle, consists of two main beams, across which are placed spruce ribs. The surface is made of Baldwin vulcanized silk of jet-black color tacked to the top of the ribs and laced to the frame. The curvature of the surface is slight and is designed for high speed. The spread is 31 feet, depth 6 feet, and surface area 186 square feet.

Direction Rudder. The direction rudder is a rectangular surface of but 6 square feet in area and is placed at the front. It is operated by wires leading to the bracket underneath the controlling column. Turning this column to either side causes the machine to turn to that side.

Elevation Rudder. The elevation rudder is likewise placed at the front and is also a single surface of 17 square feet in area. It is operated by wires leading to a lever at the side of the controlling column. Moving the column forward or backward causes the elevation rudder to turn down or up, respectively.

Transverse Control. The framework of the main plane is carried out 30 inches beyond the end of the surface on either side and affords a place for a rail on which the auxiliary sliding surfaces move. These sliding surfaces, or equalizers, are each 12½ square feet in area and when "normal" project 15 inches beyond the end of the fixed surface on either side. They are interconnected by wires, and a long cable running to each end through a pulley connects them to the steering wheel. The control is as follows: If the right end of the aeroplane is tipped down, the wheel supported on the controlling column is turned away from the lowered side. This causes the equalizer on the raised end to be pulled in under the main surface, or "reefed," while at the same time the one on the other end is pulled out. This action merely decreases the surface on the raised end and increases it on the lowered end, thus righting the machine.

Keels. A horizontal surface placed at the rear acts as a longitudinal stabilizer. It is 10.5 square feet in area and is fixed firmly to the supporting framework, 10 feet to the rear of the main surface.

Power Plant. The power plant consists of a four-cylinder, air-cooled, 25-horse-power Curtiss motor placed on the framework above the plane and to the rear of it. The motor drives direct a two-bladed wood propeller 6 feet in diameter by 4.5 feet pitch at 1,200 r. p. m. This propeller is of original design and is said to be very efficient.

General. The machine is mounted on four small, rubber-tired wheels placed at the lower ends of the four main vertical posts of the chassis. They are spaced by steel tubing and are fitted with brakes, but have no springs. The seat for the aviator is placed out in front of the main plane and directly in the center line. The total weight in flight is from 560 to 600 pounds, while the speed is estimated at 42 miles per hour; 24 pounds are lifted per horse-power, and 3.2 pounds carried per square foot of surface. The aspect ratio is 5.7 to 1.

COMPARISON OF STANDARD TYPES

From the foregoing description of what has been termed standard types, it will be apparent that, while all have many features in common, no two are exactly alike in either design, constructional detail, or efficiency. Some that are less desirable from certain points of view than other types belonging to the same class, show an unusually high degree of efficiency; others have advantages of greater stability. All, however, have proved successful in operation and some to a far greater degree than others. It will accordingly be both interesting and profitable to note the contrasts and distinctions that may be drawn. From these it will be possible to arrive at conclusions as to what particular features are most desirable at present, as well as to note what the trend of the future may be. For this purpose the aeroplanes already described are compared according to the following essential features, which are given as nearly as possible in the order of their importance, where their influence on the result aimed at—flight—is concerned: (1) transverse control; (2) aspect ratio; (3) incident angle; (4) propellers; (5) rudders; (6) keels; (7) mounting; (8) speed; (9) flight; (10) efficiency.

The object of placing the factor of efficiency last in order of importance is not to indicate that as its actual position from the practical viewpoint, as this is the one thing that designers are now striving hardest to attain, but more because it represents the best opportunity to sum up generally the performances of the different machines. Motors are compared at the conclusion of the chapter on that subject.

Transverse Control. In practice, the lateral stability of aeroplanes is maintained by four different methods: (1) automatically; (2) by warping; (3) by balancing planes, *i. e.*, wing tips or ailerons; (4) by "reefing," or the employment of supplementary sliding planes or equalizers.

At present, warping the planes is the most generally employed and most practical method, but it is expected that a simple method of automatically preserving the lateral equilibrium will be the ultimate development, and many designers, including the Wright Brothers, are striving for that end, so that it is given precedence here.

The Voisin is the only type for which automatic stability has been claimed, but it is noticeable that in later types of this machine, wing tips have been employed. The rear box-cell and the vertical keels or partitions between the surfaces of the main planes exert such a forcible "hold" on the air that to displace the machine is difficult and, in all ordinary turmoils of the air, it displays exceptional stability. In fact, a well-known aviator amusingly stated at Rheims that, were a Voisin tipped completely over on one end, it would still be aerodynamically supported, so great is the expanse of vertical surface.

Without such keels, however, the lateral balance of an aeroplane is so precarious that some form of control is absolutely necessary. The method of warping the planes in connection with the operation of the vertical or direction rudder is the chief claim of the Wright patents, and all machines employing it are essentially the same as the Wright device, even though the operating connections do not control the main planes and the rudder simultaneously. In addition to the biplanes employing it, all the successful monoplane types, except the Pfitzner, depend upon warping the main planes for this control.

Because of the structural difficulty of rigidly bracing the surface of a monoplane, warping is an ideal form of control. But the rigid structure of the biplane permits auxiliary planes to be more easily provided. This is done in the Curtiss, Farman, Sommer, the recent

Voisin, and the Cody. Both these methods of transverse control are very efficacious, but the additional resistance, unaccompanied by any increase of lift, which is produced by balancing planes, renders them less desirable than warping. On the other hand, there are objections to weakening the structure of the main surfaces by making them movable.

There is a further distinction between these two methods of control, which, although not thoroughly understood in a general sense, appears to be borne out in practice, viz, when a plane is warped the action tends not only to tip the machine up on one side but also, due to the helical form thus assumed, to turn, which can be counteracted only by the vertical rudder. In the case of wing tips, however, due to the equal but contrary position in which they are placed, both sides of the machine are equally retarded and, in addition, since the main surfaces preserve the same shape and the same angle of incidence, this tendency to turn appears to be absent. Curtiss states that to correct for tipping alone, he makes no use of the vertical rudder.

Sliding panels, or "equalizers," as applied to the Pfitzner monoplane, represent one of the recently-designed methods of transverse control which are considered not to infringe the Wright patents. This system has not been adequately tried out as yet, but there appears to be no reason why it should not be as effective as either the system of warping or the use of wing tips. There are many other methods designed to give transverse control and it seems at present that they are all equally reliable. Structural individualities of the types of aeroplanes will persist, in all likelihood, so that we can not picture the machine of the future with any one form of transverse controlling apparatus. Balancing planes and wing tips, or ailerons, are widely used at present, but further progress in aerodynamics is likely to show that warping is better, particularly as the development of improved forms of construction and more suitable materials eliminate the objection of weakening the main structure as now built.

Aspect Ratio. It is at once observable from the values given that the ratio of spread to depth (aspect ratio) of the monoplanes is generally less than that of the biplanes. This interesting fact is due very likely to the structural difficulty of making the wing of a monoplane long and narrow, and at the same time providing the necessary strength without involving undue weight. The Antoinette monoplane has recently shown a departure from this standard by decreasing the depth and increasing the spread, thus increasing the aspect ratio, but the framework had to be greatly strengthened. The new model Voisin has the highest aspect ratio of the types considered here, but exhibits no remarkable qualities therefrom.

Both theoretically and experimentally the value of this quality is considered to have much to do with the ratio of lift to drift; but whether or not in actual practice those machines like the Santos-Dumont, having as low an aspect ratio as 3 to 1, are really inferior in their qualities of dynamic support to a machine like the Cody with as high an aspect ratio as 7 to 1, is difficult to determine, since so many other factors, such as the loading and velocity, are involved. It is interesting to note here that some of the large soaring birds, such as the albatross, may be considered as aeroplanes of very high aspect ratio. The effect of aspect ratio upon speed is not discernible upon comparing the types.

Greater stability, however, is commonly supposed to result from a high aspect ratio, because of the decreased proportionate movement of the center of pressure. A further advantage is that the higher the aspect ratio of a plane, the lower is the angle giving the maximum ratio of lift to drift, and consequently for given speed and loading less power is necessary. There appears to be little question but that the development of aeroplane construction of the near future will tend toward an increase in the aspect ratio to as high, possibly, as 12 to 1.

Incident Angle. The angle that the main supporting surfaces of an aeroplane make with the horizontal line of flight is termed the incident angle, and it is something that at present varies greatly in the different types. The Wright biplane is notable for its very low angle of incidence in flight, rarely exceeding two degrees.

Renard, after deductions from the experiments of Borda, Langley, and other investigators, has enunciated the principle that, *as the incident angle diminishes, the driving power expended in sustaining a given plane in the air also diminishes.* Wilbur Wright states that, *the angle of incidence is fixed by the area, weight, and speed alone. It varies directly as the weight, and inversely as the area and speed, although not in exact ratio.* Faraud concludes that small angles are the most efficient for all aeroplanes. There is for each type a most efficient angle of incidence, or point where the power expended for flight is least. In flying, the incidence should be kept constant at this angle in order to obtain the highest speed.

The Farman, Voisin, Bleriot XI, Grade, and Sommer have an angle of incidence when first starting much greater than when in flight. Since this involves greater drift resistance and consequently more power necessary to attain the velocity of levitation and, furthermore, as aeroplanes with as heavy a loading but without an excessive angle are able to rise after a reasonably short run, it would appear as if this provision were unnecessary.

Recent experiments in aerodynamics indicate that the ratio of lift to drift, with a surface of the shape now so generally used, varies little between the values of 2 degrees and 6 degrees, a maximum value being reached in the neighborhood of 3 degrees. This explains in a measure the wide variations in this angle as observed and recorded for the different types, and also that many of the present machines preserve their equilibrium during comparatively large changes of their longitudinal inclination.

In general, the incident angle of the monoplanes is greater than that of the biplanes. The most common angle is in the neighborhood of 5 to 7 degrees. But in the Bleriot XII, an incident angle of 12 to 13 degrees is often used in flight. Incidence will very likely be established purely by the lift-drift ratio of a plane, and the angle kept as constant as possible to give this its highest value.

Propellers. With one or two exceptions, aeroplanes of all types are driven by a single, high-speed screw. The Wright and the Cody are the only instances of machines provided with two propellers rotating in opposite directions. The greater efficiency of a propeller of large diameter revolving at a slow speed over one of small diameter and high rotative speed has attracted much attention. This seems to be borne out especially in the case of the Wright machine, in which more thrust is obtained per unit of power than in any other type. The limit of rotative speed in practice is approximately 1,500 r. p. m., and in all types excepting the Wright, Cody, and Bleriot XII, the r. p. m. rate exceeds 1,000. Many of the aeroplanes, more particularly those of foreign design, use the Chauviere wood propellers, for which an efficiency of 80 per cent is claimed. The Antoinette, Grade, and Voisin, use metal screws.

The thrust and efficiency of the various propellers are about the same for equal sizes, and although the theory of propeller design is very little understood as yet, the experimental methods used have resulted in the design of propellers of as good efficiency or higher efficiency than those used in marine practice. The position of the propeller in front in most of the monoplanes is largely a matter of convenience of design, although it has an advantage in that the swiftly moving mass of air thrown backward by the screw also exerts an added lift when thrown back on the plane. At the same time, however, this action also increases the resistance, but as the frame resistance of the monoplane is much less than that of the biplane, the propeller may be placed in front without any very serious consequences. The Voisin (tractor type) biplane has the screw in front, but the results obtained indicate that this is detrimental to the speed.

It is generally believed by aviators that much better results could be obtained by the use of propellers 15 to 20 feet in diameter, rotating slowly. But there are two disadvantages involved in this feature of construction which make its adoption in the machines of the future rather doubtful. The first is the greatly added weight of so large a propeller and the second is that of building a good chassis high enough to permit of the propeller rotating freely.

Rudders. The direction rudder in all types, except the Pfitzner, is placed at the rear. The Cody biplane has an additional direction rudder in front. All the monoplanes, with the exception of the Pfitzner, have their elevating rudders at the rear, while in all the biplanes, except the new Wright and more recent Voisin models, this rudder is placed out in front. Rudders placed at the rear are advantageous in that they act at the same time as keels. But, in general, the placing of the elevating rudder in front seems to offer more exact control of the longitudinal stability.

The elevating rudder almost always exerts some supporting power. Therefore, when placed in front and turned up for ascent, the support is increased, as it naturally should be. But when this rudder is placed at the rear, the movement for ascent is such that the supporting power of the rudder is decreased, making it of negative value, so that instead of causing the front of the machine to rise, it causes the rear to sink. Following the same theory shows that when the elevating rudder is out in front, in starting, the front of the machine lifts off the ground and is strongly followed by the body; while if this rudder be in the rear, when turned to give ascent the rear merely sinks more, and not only is the run greatly increased, but the power required and the risks incurred are greater. That it is generally so used on the monoplanes is the result of necessity due to the propeller being at the front.

In the Wright biplane the elevating rudder is so constructed that when elevated it is automatically warped concavely on the under side, and when depressed, curved in the opposite way. This materially adds to the rudder's force due to the peculiar law of aerodynamics whereby a curved surface, under the same conditions as a flat surface, has a greater ratio of lift to drift. The reduction in the size of the rudder is thus made possible and its flat shape when normal greatly reduces the head resistance. In so far as a biplane is usually supposed to cause interference of the two surfaces and greater head resistance, it would appear as if the biplane rudders as used on the Wright and the Curtiss were not as efficient as single planes, but the structural advantages of this arrangement are important.

The method employed by Grade of merely bending flexible surfaces, instead of turning rigid movable planes, has a great advantage in that the rudders, after being used, spring back to their normal position. This method has not been adopted on any other type, however, although it has many considerations of safety favoring it.

In almost all of the successful aeroplanes, with the possible exception of the Wright and Antoinette, it is conceded that the size of the rudders is much too great. This is clearly indicated by the remarkably small change of inclination usually necessary for a change of direction. This ultra sensitiveness where, as in some machines, a movement of a few hundredths of an inch will considerably alter the state of equilibrium of the machine, is certainly undesirable. To begin with, it need hardly be pointed out that oversensitiveness of a rudder usually invites dangerous situations. Furthermore, if a rudder be extremely sensitive, it is a good indication that it is too large, in which case it is absorbing considerable power that could be put to better use elsewhere. It is quite likely, therefore, that a great decrease in the size of the rudders will be a development of the near future.

Keels. Keels on aeroplanes, like keels on a boat, add greatly to the stability. But on an aeroplane they are "dead surfaces" and, as such, have the disadvantage of offering greater expanse of surface for wind disturbance to act upon. They unquestionably decrease the speed. Tapering keels, such as those employed on the Antoinette, the Pelterie, and the latest Bleriot XI, offer a maximum of "entering edge" with a minimum of area, and for that reason are more advantageous than those of rectangular form. Keels are entirely lacking in the original Wright, Santos-Dumont, and Cody, but in the later "headless" Wright two small, vertical keels of semicircular form are placed in the angles made by the meeting of the skids with the braces from the latter to the upper main plane, while a horizontal keel of considerable area is employed in the rear of the Short Wright (English manufacture).

In the Voisin, use is made of several vertical keels of large area, really partitions, placed not only in the rear cell but also between the main planes themselves. That these have not proved entirely satisfactory is indicated by the adoption of ailerons to maintain transverse stability in the more recent Voisin machines. Keels add greatly to the resistance of a machine, the head resistance and skin friction with their consequent power absorption being considerable. It is generally conceded now that control by rudders is becoming so perfected that any inherent stability to be obtained by the use of keels at the expense of power is hardly worth while. No special form or combination of keels, so far designed and tried, have really succeeded in giving any kind of complete inherent stability.

Actual practice, however, demonstrates that they do increase stability, tending to hold the machine to its course, and keels at the rear of a machine somewhat on the order of a bird's tail are found advantageous, so that it is quite unlikely that they will disappear as a feature of aeroplane design for years to come.

Mounting. This is the only remaining detail of construction that need be considered in this connection. There is probably no other single feature in which the various machines differ more widely, nor any other in which such totally different provision, or the absence of it, for absorbing the shocks of landing, proves so uniformly successful. When an aeroplane drops as a dead weight for even a short distance, it suffers considerable damage regardless of the presence of shock absorbers or otherwise, whereas, in ordinary use, it appears to be as easy to land lightly with a machine having a rigid chassis, as with one in which elaborate precaution is taken to guard against shocks.

There is another factor to be guarded against, however, and that is the gyroscopic action developed by the swiftly revolving propeller, which tends to resist a sudden change of its plane of rotation, as well as all vibration. If, therefore, when running over the ground the machine be suddenly jarred, the propeller is likely to snap off. This has been experienced by M. Bleriot on more than one occasion, and he emphasizes the necessity of providing springs on a heavy machine mounted on wheels.

Three distinctive methods of mounting have been employed to date:

(1) Using skids only, as in the original Wright machine. This is already obsolete, as it involved the use of special starting apparatus.
(2) Wheels only, as in the Curtiss, Voisin (both types), Bleriot (both types), Pelterie, Grade, and Pfitzner.
(3) Wheels and skids combined — Farman, Antoinette, Santos-Dumont, Cody, Sommer, and later Wright machines.

Details of some of the most important designs are given in Fig. 40.

The relative merits of mounting on wheels only or skids and wheels constitute a subject of wide discussion. Where the former are employed independently, the addition of a brake is almost indispensable to bring the machine to a quick stop where the landing area is restricted; whereas, with a skid forming part of the support, as in the Antoinette, the latter acts as a brake. Of course, it performs the same office in starting, to the detriment of a quick rise from the ground. The extra power required on that account, however, is not very great, as the skid, supporting only the tail, does not carry any great weight. It is consequently not very efficient as a brake so that provision of the latter class on all types should be made.

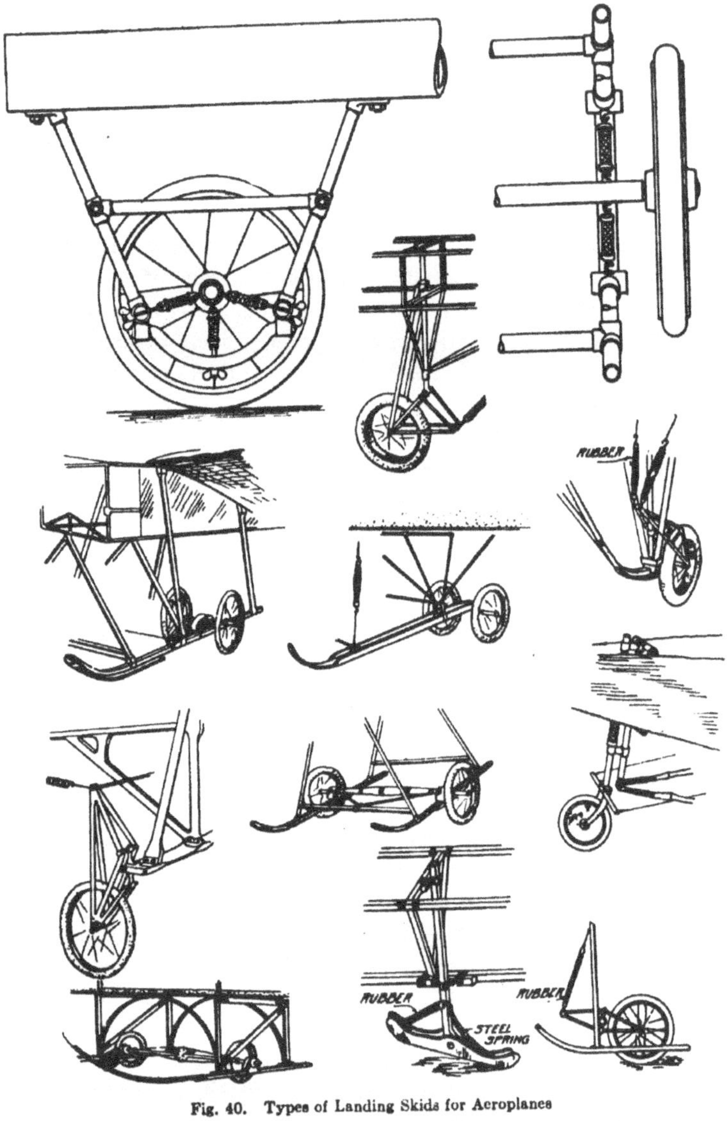

Fig. 40. Types of Landing Skids for Aeroplanes

A number of combinations of skids and wheels have been tried, such as that of the new Wright which starts on its four wheels, and lands on the skids to which they are attached. The Sommer and Farman are typical examples of this combined form of mounting, and experience in their use appears to demonstrate that they are the most effective for heavy machines. On light aeroplanes, such as the Curtiss and the Grade, where the loading is reasonably light, spring mountings have been found unnecessary, the wheels alone taking the shock of landing. No skids are employed. The more recent Curtiss machines are provided with an efficient brake. It is quite likely, however, that the high speed aeroplane of the future will not only be provided with an elastic mounting, but when regular stations are established, means will be employed for projecting it from some ingenious starting device at high velocity, so that it may be quickly launched into the air.

Speed. It is generally conceded that the chief object of the aeroplane designer at present is to increase the speed, prophecies of 100 miles per hour, and considerably more, being not at all uncommon. Whether this can be attained or not is a question that only the future can solve, but a comparison of the speeds prevailing in January, 1910, and December, 1910, shows such a marked increase for the development of a single year that this does not appear to be beyond the possibilities of the future, by any means. It must be borne in mind, of course, that while resistance increases as the square of the speed, the power to overcome it must increase as the cube. This would seem to make the attainment of the 100-mile mark something that would involve considerable modification of the present type of aeroplane, in order to attain increased efficiency, as the goal in view is not to be won by a mere increase of power.

The speed shows no direct variation with aspect ratio or loading, and higher speeds seem to be merely attained by an excess of power, a decrease of head resistance, and a small supporting surface. In Table I are given the speeds of the various types described, *i. e.*, those of which they had been shown capable up to January, 1910.

Flight. In the manner of flight of the different types, pronounced distinctions may be drawn. Probably the widest variation in this respect exists between the Wright and the earlier Voisin

TABLE I

Speed Data

Type	Miles per hour	Type	Miles per hour
Bleriot XI (Racing Type)	63	Farman (Racing Type)	44
Santos-Dumont	55	Sommer (1910 Model)	44
Bleriot XI Bis(1910 Model)	51	Wright (Rear Control)	43
Antoinette (1910 Model)	50	Wright(1910 Model)	41
Voisin (Racing Type)	49	Farman (1910 Model)	41
Curtiss	48	Voisin (1910 Model)	40
Bleriot XII (1910 Model)	48	Farman (Passenger Type)	39
Bleriot XI 2 Bis	48	Pelteric	39
Pfitzner	45	Cody	37
Grade	44		

with numerous vertical keels. The flight of the latter may be best described as "sluggish." The enormous resistance of this machine appears to hold it back very perceptibly, while in making turns its action is slow and "deadened." In sharp contrast to this is the strikingly active flight of the Wright machine. Its resistance is very small for a biplane and its movement through the air is quick and precise, particularly when compared with the flight of the Farman biplane, the sluggish movements of which at the International Meet near New York earned for it the sobriquet of "the ice-wagon." In changing direction or warping to maintain lateral stability, the action of the Wright is precise and almost instantaneous, the Wright biplane answering its helm in a remarkably quick and effective manner.

In grace of form and swiftness of flight the Antoinette and Bleriot monoplanes are a delight to the eye. They appear to move through the air without the slightest effort and at the distance of a mile or so give the impression of being huge, soaring birds, so steady and perfectly under control is their every movement. Due to the smooth whirring of their multi-cylindered motors, this is accentuated when close at hand, in comparison with the clattering exhaust of the four-cylinder Wright engine, which many uninitiated spectators mistake for a noise made by the propellers, the turning of which is plainly visible owing to their low speed.

The Curtiss in flight is noticeable for its constant rising and

TABLE II

Characteristics of Different Types

Machine	Pounds per h. p.	Pounds per sq. ft.	Speed in Still Air, m. p. h.
Wright	41	2.05	41
Wright (r. c.)	37	2.50	43
Farman (pssgr)	34	3.15	39
Bleriot XI	29	4.50	51
Antoinette	27	3.33	50
Pelterie	27	4.40	39
Cody	25	2.57	37
Farman ('10)	24	2.80	41
Pfitzner	24	3.20	45
Curtiss (pssgr)	22.6	3.64	..
Voisin ('10)	22.5	3.14	40
Curtiss	22	2.50	48
Farman (rcg)	21	3.00	44
Bleriot XII	21	5.30	48
Bleriot XI 2	19.8	3.68	48
Voisin (rcg)	19.5	3.27	49
Voisin (tractor)	19	2.36	40
Grade	17	2.00	44
Sommer	16	2.76	44
Sommer (rcg)	15	3.25	..
Santos-Dumont	12	3.10	55
Bleriot XI (rcg)	7.5	5.76	63

falling, tracing a sinuous, vertical path through the air, in contrast with the perfectly even and level keel maintained by most of the other machines. In making any comparison of flight, however, the personal equation also must be considered, as the action of the machine is largely governed by the skill of the aviator. In the present instance, however, the impressions recorded were of the different machines in the hands of skillful pilots, all of whom had been flying for a year or more and had made a great many flights. The Santos-Dumont was early dubbed the "clown" of the International Meet and its appearance was invariably the signal for a roar of amused applause. Despite its speed, its erratic action marked by continual dips and violent rocking, always seemed to have it on the verge of tumbling to the ground. Because of its light loading, the Grade seems especially buoyant in the air. The other types mentioned show characteristics between the extreme sluggishness of the Voisin and Farman and the remarkable preciseness of the Wright.

As the question of duration of flight depends much more upon the skill of the aviator, the endurance of the motor, and the amount of fuel carried than it does on the machine itself, a comparison of the longest flights made by each type would be valueless.

Efficiency. One of the best indications of the general efficiency of an aeroplane is the amount of weight carried per horse-power, but it will be apparent that this must also be considered in connection with the weight lifted per square foot of lifting surface, its speed, and similar factors. The first of these mentioned is usually termed "pounds per horse-power" and is obtained by dividing the total weight of the machine in flight by the horse-power of the motor. This is a variable owing to the different weights of the aviators, but not one of sufficient importance to record in the case of a one-man machine. At the time Table II was compiled (January, 1910), the Bleriot XI (racing model) appeared to be the most wasteful of power, while the Wright was the most efficient, this still being true of the latter. It must also be borne in mind that the Bleriot is a very much faster machine than the Wright. Table II summarizes the various characteristics of the different types.

EXAMINATION PAPER

TYPES OF AEROPLANES

PART I

Read Carefully: Place your name and full address **at the head of the paper. Any cheap, light paper like the sample previously sent you may be used. Do not crowd your work, but arrange it neatly and legibly.** ***Do not*** *copy the answers from the Instruction Paper; use* ***your own words, so that we*** *may be sure you understand the subject.*

---

1. Enumerate the most prominent types, classifying them into biplanes and monoplanes.

2. Give a definition of "transverse control" in your own words.

3. What is meant by *aspect ratio?*

4. What material is used in the frame construction of the Wright biplane, Model B?

5. Give the dimensions of the supporting planes of the same machine.

6. Give a clear description of the transverse control of Wright's biplane.

7. Draw a side elevation and plan of Wright's Model B.

8. How is the transverse control effected in the Curtiss blpiane?

9. Make a side elevation of Voisin's biplane.

10. How is the elevation rudder constructed in the Voisin?

11. Make a sketch and give some characteristics of the Farman biplane.

12. Make side elevation of Cody's biplane.

13. Give general data about Cody's machine regarding weight, speed, efficiency, and aspect ratio.

14. Give a brief general description of Antoinette's monoplane.

15. What is the most distinctive feature of Santos-Dumont's Demoiselle?

16. Sketch an elevation of Bleriot XI monoplane.

17. Describe the supporting planes of this machine.

18. What changes have been made in later types of this monoplane?

19. What are the general features of the Grade monoplane?

20. How is the elevation controlled with the Pelterie monoplane?

21. In what general arrangements does the Pfitzner monoplane differ from others?

22. By what methods is the lateral stability of aeroplanes maintained?

23. What is the chief claim of the Wrights' patents?

24. How is it supposed that the stability changes with the aspect ratio?

25. What is the *incident angle?*

26. Where are the direction rudder and elevation rudder mostly placed?

27. For what reasons are keels employed?

After completing the work, add and sign the following statement:
I hereby certify that the above work is entirely my own.

(Signed)

Spotted something that isn't quite right?

Let us know and we'll fix it for future reprints. And as a thank you for your diligence, we'll send you a free Sothis book of your choice.

publisher@sothis.press

This page intentionally left blank

Except of course for that line above. And this one.
And whilst we have your attention, perhaps you'd
like to follow us on Instagram @sothis.press. Go on.

Made in United States
Cleveland, OH
09 November 2026

47955553R00052